W9-AOE-434

DATE DUE

THE TWO-MILE TIME MACHINE

THE TWO-MILE TIME MACHINE

PRINCETON UNIVERSITY PRESS   Princeton and Oxford

Richard B. Alley

THE TWO-MILE TIME MACHINE

Ice cores,

abrupt

climate

change,

and our

future

Copyright © 2000 by Princeton University Press
Published by Princeton University Press, 41 William Street,
Princeton, New Jersey 08540
In the United Kingdom: Princeton University Press,
3 Market Place, Woodstock, Oxfordshire OX20 1SY

Library of Congress Cataloging-in-Publication Data

Alley, Richard B.
The two-mile time machine : ice cores, abrupt climate change, and our
future / Richard B. Alley.
p.   cm.
Includes bibliographical references and index.
ISBN 0-691-00493-5 (alk. paper)
1. Paleoclimatology.   2. Climatic changes.   3. Ice—Greenland—
Analysis.   I. Title.

QC884.A55 2000
551.6'09'01—dc21   00-036730

This book has been composed in ITC Garamond Light with Gill Sans display.

The paper used in this publication meets the minimum requirements of
ANSI/NISO Z39.48-1992 (R 1997) (*Permanence of Paper*)

www.pup.princeton.edu

Printed in the United States of America

10   9   8   7   6   5   4   3

CONTENTS

CONTENTS

SETTING THE STAGE

Why we might care what happened to Earth's climate in the past, and what might happen in the future

We live with familiar weather—ski areas are snowy, deserts are parched, rain forests drip. But what if our climate jumped to something totally unexpected? What if you went to bed in slushy Chicago, but woke up with Atlanta's mild weather? Or worse, what if your weather jumped back and forth between that of Chicago and Atlanta: a few years cold, a few years hot? Such crazy climates would not doom humanity, but they could pose the most momentous physical challenge we have ever faced, with widespread crop failures and social disruption.

Large, rapid, and widespread climate changes were common on Earth for most of the time for which we have good records, but were absent during the few critical millennia when humans developed agriculture and industry. While our ancestors were spearing woolly mammoths and painting cave walls, the climate was wobbling wildly. A few centuries of warm, wet, calm climate alternated with a few centuries of cold, dry, windy weather. The climate jumped between cold and warm not over centuries, but in as little as a single year. Often, conditions "flickered" back and forth between cold and warm for a few decades before settling down.

The history of this climatic craziness is written in cave formations, ocean and lake sediments, and other places. But the record is probably clearest and most convincing in the ice

3

of Greenland. This incomparable, 110,000-year archive provides year-by-year records of how cold and snowy Greenland was, how strong the storms were that blew dust from Asia and salt from the ocean, and even how extensive the wetlands of the world were.

These records show clearly that Earth's climate normally involves larger, faster, more widespread climate changes than any experienced by industrial or agricultural humans. The 110,000 years of history in Greenland ice cores tell of a 90,000-year slide from a warm time much like ours into the cold, dry, windy conditions of a global ice age, a 10,000-year climb back to warmth, and the 10,000 years of the modern warm period. But the ice cores also show that the ice age came and went in a drunken stagger, punctuated by dozens of abrupt warmings and coolings. The best known of the abrupt climate changes, the Younger Dryas event, nearly returned Earth to ice-age conditions after the cold seemed to be in full retreat. The Younger Dryas ended about 11,500 years ago, when Greenland warmed about 15°F in a decade or less. A little more, slower warming then led to our current 10,000 years of climate stability, agriculture, and industry.

But smaller and slower climate changes during recent millennia have affected human civilizations in many ways—and these small climate changes seem to have been getting bigger. The "Little Ice Age" cooling that changed settlement patterns in Europe a few centuries ago was tiny compared to the Younger Dryas or the global ice age, but seems to have been the biggest change for thousands of years.

Records from many places beyond Greenland provide a longer, if fuzzier, view of climate history. Over the last million years, the pattern recorded in cores of Greenland ice has occurred over and over: a long stagger into an ice age, a faster stagger out of the ice age, a few millennia of stability, repeat. The current stable interval is among the longest in the record. Nature is thus likely to end our friendly climate, perhaps quite soon; the Little Ice Age may have been the first unsteady step down that path.

In our climate, great ocean currents sweep north along

the surface of the Atlantic, are warmed by the tropical sun, and release that heat into the winters of northern Europe, allowing Europeans to grow roses farther north than Canadians meet polar bears. The ocean waters that cool in the north Atlantic then sink into the deep ocean and flow south on the first stage of a globe-girdling journey before returning. This "conveyor belt" circulation is delicately balanced—add a little too much fresh water to the north Atlantic from rain or melting icebergs, and the wintertime ocean surface will freeze to produce floating sea ice rather than sinking to make room for more hot water. Much evidence shows that the abrupt coolings and warmings occurred when the conveyor circulation suddenly shut off or turned on again, triggering other changes that spread across Earth.

Human-induced greenhouse warming appears capable of triggering a conveyor shutdown, by increasing precipitation in the far north and by melting some of the remaining ice sheet on Greenland. Strange as it seems, "global warming" may actually freeze some regions! But, if we slow down the warming, it is just possible that we can avoid an abrupt change and even help stabilize the climate.

This book is a progress report on abrupt climate changes. We will discuss what has been learned, how this knowledge was gained, and what it might mean to us. The existence of abrupt climate changes casts a very different light on the debate about global warming, so we will examine the greenhouse arguments under this new light. We won't find all of the answers—many are not known yet—but we will frame the questions, and we may gain some clues to our future.

Climate Matters

Climate matters. It mattered to the Vikings, who settled Iceland, explored the New World, and were lured north to Greenland during a period of unusually warm weather a millennium ago. But the warmth did not last, and Viking settlements on Greenland slowly contracted as the climate cooled into the Little Ice Age (see Figure 1.1). The settlers brought farm animals into their houses during the cold winters. Eventually, the settlers ate their farm animals, then their dogs, then disappeared themselves. Climate mattered to Oklahoma farmers during the Dust Bowl years of the 1930s, when many people headed west as much of their soil headed east on withering winds. Today, with floods and drought, feast and famine, climate matters to many of us much of the time.

To be fair, climate is not everything. The victims of the Dust Bowl and of the cooling in Greenland may have contributed to their own plights through farming practices that pro-

FIGURE 1.1
The history of temperature and the rate at which snow accumulated in central Greenland over the last ten thousand years, reconstructed using techniques we will discuss soon. The horizontal bars indicate the mild, wet Medieval Warm Period, when the Vikings settled Greenland, and the colder, drier Little Ice Age, which helped drive the Vikings from Greenland. These records have been "smoothed" by averaging over a century or so, making a short cold time about 8200 years ago appear smaller than it was. The 1 to 2-degree shifts shown here are the kinds of climate change that most experts worry about. The data are from the 1997 paper by Cuffey and Clow cited in the Sources and Related Information.

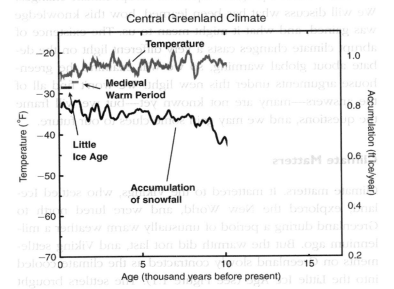

Central Greenland Climate

moted soil erosion, and the Oklahomans were fleeing a great economic depression as well as a change in the weather. While the Vikings froze out of Greenland, their "Eskimo" neighbors, the Thule Inuit, readily survived the cooling.

Still, the Assyrians, the Maya, the Anasazi, and other ancient civilizations seem to have risen to glory while nature watered their crops, and to have fallen when those crops dried out. Climate certainly mattered to them, and it certainly will matter to us.

One of the important debates of our time centers on global warming. On one side are those who argue that human-caused changes in climate will make our lives so difficult that millions of us may die, and the fabric of our civilization may be changed forever. On the other side are those who warn that efforts to avoid such a hypothetical fate may cause us to commit economic suicide and trigger the decay that we fear. To resolve this important debate, thousands of people and millions of dollars are currently devoted to the development of an "Operator's Manual" for planet Earth. Land and water, air and ice, soils and plants—if we can figure out how they work, how they are wired together and depend on each other, maybe we can then make wise decisions about global warming, ozone depletion, and other globe-girdling questions.

This effort is called Earth system science. It is mostly about observing Earth here and now, understanding modern processes, and building models of those processes to use in making predictions. But history also plays a role, in two ways. Just as the records of past peoples help us understand human society, the records of past climates help us learn how the Earth system works. And just as modern political scientists can test their ideas against the history of humans, Earth system scientists can test their models against past climate changes.

The climate models these scientists test are highly altered, computerized weather forecasting tools. If you decide to learn to forecast the weather, every day offers a new problem, and the next day provides the answer in the back of the book. A weather forecaster in training can practice predicting tomorrow's weather more than a thousand times during a college career.

Forecasting the climate is not so easy. Consider a hypothetical modeler who informs a U.S. congressional committee that disaster will arrive in a century unless we change our ways. The chair of the congressional committee is unlikely to subscribe to the doctrine of scientific infallibility, and may harrumph that economic policy should not be based on untested computer output. The real winners and losers of such a debate will be the great-grandchildren of the disputants, because modeler and congressperson alike will have been recycled themselves before the forecast can be tested.

It would be better if the scientist could also tell the congressional committee, "The model that predicts future problems has been tested by simulating many climates of the past, that were wetter and drier, warmer and colder, with greenhouse gases higher and lower than today, and the model successfully reproduced the observations. The model has been used to run simulations that started in the previous warm period and went through the most recent ice age to today, and successfully matched the changes that brought us here." The congressional committee would have a much harder time dismissing such a scientist as a crackpot. But to test our models against the history of climate, we must know that history.

These questions are far from academic. The Medieval Warm Period that opened Iceland, Greenland, and North America to the Vikings, and the Little Ice Age cooling that helped drive the Vikings from Greenland, caused glaciers to advance across farms in Norway, and allowed Hans Brinker to skate on the canals of Holland, were dwarfed by the Younger Dryas and other dramatic climate jumps that ended the last ice age, as shown in Figure 1.2. Some climate models suggest that such jumps could return, and that human activities may cause—or prevent—such a return. Many of us believe that it would be prudent to understand the large climate jumps of the past. This book is an attempt to advance that understanding in some small way.

In the next chapter, you will find a brief introduction to climate history, including the central role played by ice cores. I have been fortunate to help in reading ice-core records during three trips to Antarctica, five trips to Greenland, and countless hours in frozen laboratories. Most of us who study ice cores started out by trying to learn how the ice actually records climate, and Part II of this book provides an introduction to the many methods we use. These methods have taught us amazing things about the climate, which are described in Part III. Those results have forced us to learn about oceanic and atmospheric processes far beyond the ice sheets, as described in Part IV. Finally, all this effort gives us some insight to the future, as described in Part V.

FIGURE 1.2

The history of temperature and the rate at which snow accumulated in central Greenland over the last 17,000 years. The prominent Younger Dryas cold event, and the warmings and coolings before it, dwarf the climate changes that helped chase the Vikings around. The scales on this figure are identical to those on Figure 1.1, and the data are from the same source.

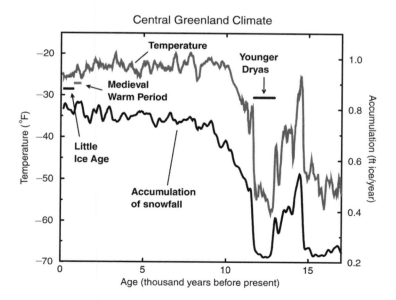

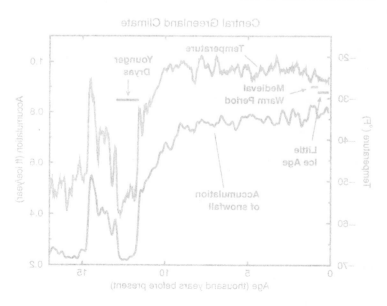

2

To read the record of past climate shifts, we have to find the right history book. Humans hadn't yet mastered writing the last time the climate jumped, so we can't look up the answer in the library. Fortunately, there is a sort of "library" in ice sheets, lake beds, and the ocean floor that tells us much of what we wish to know.

Archaeologists poke around in the trash dumps of ancient humans, looking for lifestyle clues. "Modern archaeologists" do the same in modern trash dumps, sorting out the Barbie-doll heads from the half-eaten hot dogs in our garbage to learn perhaps more about us than we wish they knew. The "garbage" of the Earth system is called *sediment,* and it piles up in many places. Common sense and careful study allow us the "read" the records in that sediment, learning when it was deposited, and what the world was like then.

Sediment itself provides clues to its origins. A glacier scratches and polishes the stones it drags over other rocks, while the wind sandblasts the grains it piles into dunes, and mud settles in regular layers in lakes. We can easily recognize a cold climate from its glacial deposits, a dry climate from the sand dunes it leaves, or a wet climate from lake sediment.

If you sift through the lake sediment, you will usually find many other interesting things. Windblown pollen is readily identified in old sediment, and pollen from sagebrush, palm trees, or tundra flowers will tell very different stories of how dry or wet, hot or cold the climate was around the lake. Creatures living in many lakes leave their shells in the sediments, but different kinds of creatures with different shells live in salty or fresh, warm or cold lakes. Many other indicators exist, and we will discuss these indicators as they become important to our story.

Most paleoclimatologists spend their time looking at ocean sediments. The oceans cover more than two-thirds of the planet, and sediments accumulate in them almost everywhere. In comparison, lakes and sand dunes are rare, with most of the land surface slowly being washed or blown away rather than being buried in sediment. Paradoxically, it is also easier to study sediments from the ocean than from the continents. Large, specially equipped drill ships dedicated to collecting sediment cores from the ocean floor are operating year-round. But coring a small pond may involve trying to find a drill, trying to find a boat, persuading the landowner to let you in, getting drill and boat through the mud around the pond, falling out of the boat while trying to recover some mud from the bottom, trying to come back when the lake is frozen to allow coring from the ice, getting stuck in a snowdrift, and so on.

Fortunately, there are many people who seem to enjoy mud and mosquitoes, or slipping on snow, so lakes are being cored. Other records are being collected from cave formations, tree rings, and even the odds and ends that pack rats have gathered around their dens and then urinated on, preserving for posterity.

Perhaps the finest records of past climate are obtained from glaciers and ice sheets. Fully one-tenth of Earth's land surface is buried by ice, mostly in the vast Antarctic and Greenland ice sheets, but also in thousands of smaller mountain glaciers. The great ice sheets are more than two miles thick in the middle, and have been accumulating snow, and climate records, for hundreds of thousands or even millions of years. Climate scientists use drills to cut into ice sheets and glaciers, collecting ice cores. An ice core is a cylinder of ice, typically four or five inches across. Cores often are collected in three-foot-long sections, but when placed end to end, these sections can extend for two miles and include the entire history of the ice sheet and the air above it.

As we will discuss soon, ice cores reveal past temperatures, and tell how much it snowed, how windy it was, how often forest fires occurred upwind, how productive the surrounding oceans were, how active the sun was, and even how much carbon dioxide was in the air and how widespread wetlands were on Earth. The ice brings all this information together, providing a surprisingly complete history of climate changes over much of Earth's surface.

Interpretation of ice cores, and of many other climate records, has recently revolutionized our view of Earth. We once believed that the climate is well-behaved—a little change in the brightness of the sun, or the positions of the continents, or the composition of the air causes a little change in the climate. The ice cores tell a more complicated story. Sometimes, a small "push" has caused the climate to change a little, but other times, a small push has knocked Earth's climate system into a different mode of operation, bringing new weather patterns to much of Earth in only a few years or decades. To scientists accustomed to changes over geological time, it is almost as if someone had flipped a switch to change the climate. Sometimes, the climate jumped back and forth a few times before settling into one pattern, almost as if the person flipping the switch were an impish three-year-old. The climate jumps have been much larger, quicker, and more widespread than those that chased the Greenland Vikings and the Oklahoma farmers from their homes, or those experienced by any other agricultural or industrial humans. Were such changes to occur today, the consequences could be severe.

We do not yet know how to predict such changes, or even how likely they are to recur. But we are learning many of the key places to study, many of the pieces of the puzzle, and the right questions to ask. So, let's go to Greenland and get started.

READING THE RECORD

How we learn what happened to Earth's climate in the past

3

Drilling ice cores in Greenland or Antarctica usually involves getting on a ski-equipped plane and flying a few hundred miles over snow to a place where people live in tents at a temperature of thirty degrees below zero. Using an assortment of snowmobiles, skis, caterpillar tractors, computers, and a lot of "elbow grease" and "bigger hammers," drillers race the short summer to pull sticks of ice up from the depths beneath their feet, analyze that ice, and ship it home. In the next few pages, we will take a quick look at the origins and practice of this odd pastime.

A Brief History of Ice Coring

No one knows whether the "Ice Man" noticed the layers in the crevasse walls of the Alpine glaciers he crossed 5,000 years ago, before he fell in and was frozen until his body was discovered recently. Nor do we know who first conceived of coring into such glaciers to study those layers. Many observers credit Henri Bader, of the U.S. Army Corps of Engineers, with initiating the modern era of ice-core drilling that ultimately led us to Greenland.

Armies have long been interested in the cold. Bitter winters helped drive Napoleon from Moscow, and have tested soldiers and their equipment throughout history. The U.S.

Army Corps of Engineers has an institute to help its people and equipment deal with the cold, which is now known as the Cold Regions Research and Engineering Laboratory (CRREL) and is located in Hanover, New Hampshire. This laboratory is a center of expertise in fundamental and applied research on snow, ice, sea ice, frozen ground, and more. Civilians have undoubtedly benefited more from research by CRREL than has the military, and a surprising amount of what we know about cold places can be traced to CRREL's efforts.

When the awakening realization of global interdependence led to the International Geophysical Year of 1957–58, much of the research was focused on the little-explored cold regions of Greenland (see Figures 3.1 and 3.2) and Antarctica. Henri Bader, in his role as chief scientist of the major U.S. lab, promoted coring through glaciers for scientific study. Development projects were begun then, and carried forward vigorously. By 1966, CRREL had cored most of a mile through the far northwest corner of the Greenland ice sheet at Camp Century. Two years later, Tony Gow was analyzing more than a mile of ice from the first core through the Antarctic ice sheet

FIGURE 3.1
During the GISP2 drilling, this fox trotted into camp, having crossed more than one hundred miles of ice sheet with no apparent harm.

at Byrd Station. These efforts were followed by other activities by CRREL and others, including drilling through the Ross Ice Shelf in Antarctica, and the Danish-Swiss-U.S. Greenland Ice Sheet Project (GISP), which completed a core through the southern dome of Greenland at Dye 3 in 1981.

The scientific return from these early efforts was outstanding. Pioneering polar researchers added the third dimension to the great white blobs of ice on their maps, and showed how the ice flowed and changed. Careful analyses demonstrated that the ice has been piling up, layer on layer, for at least tens of thousands of years, and that this ice contains a record of the climates that produced it. Tantalizing clues suggested that those records would change the way we view our world.

However, the early coring was limited by its immense difficulty. The cores usually were drilled where camps already existed (as seen in Figure 3.3), not necessarily in the most

FIGURE 3.2
The interesting "sun dogs" are the sun reflecting off ice crystals growing in the air during the colder "night," when the sun is lower.

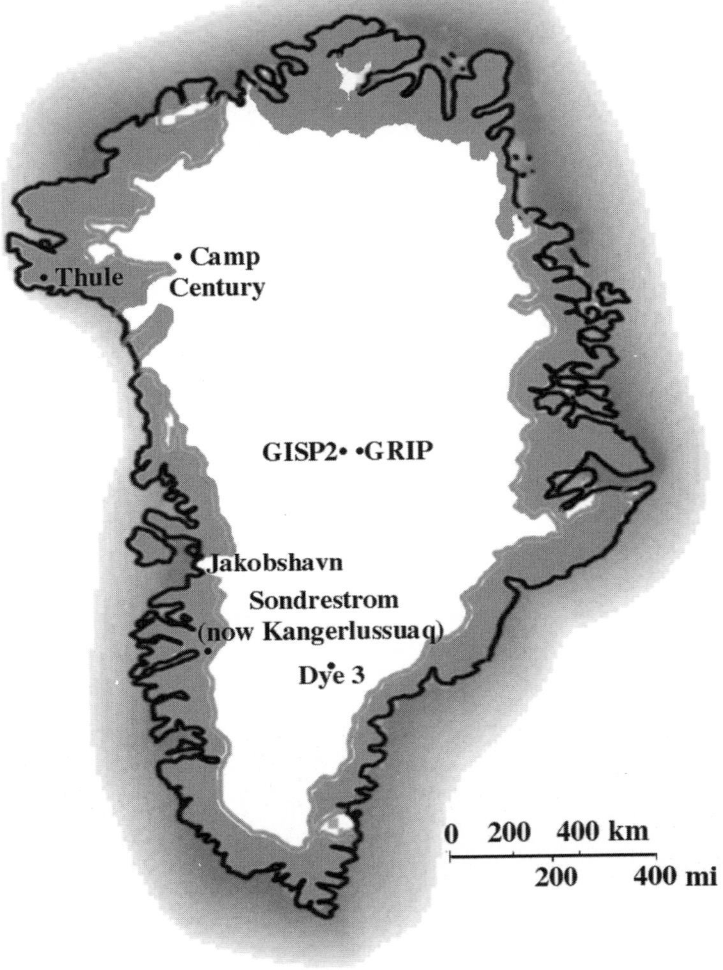

FIGURE 3.3
Map of Greenland, showing the sites where deep ice cores have
been collected, the coastal stations Sondrestrom (now Kanger-
lussuaq) and Thule used to support the ice-coring operations, and
the Jakobshavn glacier, which maintains a higher ice-flow speed
than any other glacier on Earth.

interesting places. The ice-core records from Greenland were especially fascinating, for reasons that will become apparent soon. Unfortunately, the really startling stories were in ice from very close to the bottom, where the flow of the ice sheet had made the records very hard to read and understand.

As concern built about climate change and global warming, many people raised the call for new Greenland ice cores in the right places to answer the big questions. Collecting two cores close to each other would show whether the records were reproducible, and thus believable. International cooperation would improve the quality of the science. Plans were made, agreements were signed, permits were obtained, and drills were designed.

This intense activity produced the GISP2 and GRIP projects. GRIP, the Greenland Ice Core Project, was primarily a European consortium, and drilled through the summit of the ice cap between 1989 and 1992. GISP2, the Greenland Ice Sheet Project 2, was primarily a U.S. effort, and drilled through the ice twenty miles west of GRIP between 1989 and 1993.

While all this planning, building, and surveying were going on, I was a still a student, working my way through school at Ohio State and then Wisconsin. In 1985, I spent a few weeks in Greenland a bit south of the site later chosen for the GRIP camp. There, I helped drill and analyze one of the short, 300-foot ice cores that guided us to the best sites for the deep coring. As I graduated from student to professor, I was also asking for permission and money to join the GISP2 project. Paul Mayewski, the prominent ice chemist at the University of New Hampshire, was in charge of the science. A distinguished team of researchers was being assembled, including Tony Gow, CRREL's ageless guru of ice who had been basically the entire science team during the Byrd Station drilling twenty years earlier. It was my good fortune to be selected to work on some of the analyses with Tony and his colleague, Deb Meese.

Very briefly, each of the five years of GISP2 included work that spanned April to September. Staff went in early to build or open camp, followed by the main invasion of scien-

FIGURE 3.4
The author on the Greenland ice sheet.

tists (Figure 3.4). The science effort was typically split into two six-week-long "legs." For studying the physical properties of the ice core, my research team, along with Joan Fitzpatrick of the United States Geological Survey, handled one leg each summer, and Tony Gow, Deb Meese, and coworkers handled the other. After the second science leg, a short period of frantic packing and cleanup left the site ready for the next year. Each leg featured about fifty people on-site, to handle drilling, science, cooking, unloading and loading the ski-equipped airplanes that supplied us, and more (see Figure 3.5).

Doing the Job

The GISP2 camp was a carefully choreographed dance of several partners. The U.S. National Science Foundation funded GISP2, with the European Science Foundation funding the parallel GRIP effort. The U.S. 109th Air National Guard of Scotia, New York, provided the heavy-lift aircraft that supported both camps. The 109th's ability to deliver tons of supplies to an ice sheet safely and reliably has revolutionized polar research.

The Polar Ice Coring Office (PICO) was contracted to drill the GISP2 core and run the camp. PICO has been involved in polar drilling and logistics for a couple of decades, an outgrowth of an Antarctic program to drill through the Ross Ice Shelf. Interface with the scientists, core processing, and a host of other key jobs were handled by the Science Management Office of the University of New Hampshire, directed by Chief Scientist Paul Mayewski. When a big project succeeds, you can almost guarantee that some "hero" made it happen by solving each of the myriad problems that arise over the long years of planning and doing. Paul Mayewski was that hero for GISP2.

One of the difficulties of GISP2 was figuring out how to analyze the ice. Today, the United States has a wonderful National Ice Core Laboratory in Denver, directed by our colleague Joan Fitzpatrick. Ice cores are analyzed as well as archived there. But when GISP2 began, an appropriate U.S.-based lab was not available. So it was decided to build a lab in the ice sheet, and do the work there.

This was not nearly as easy as it sounds. The analyses included electrical, visible, and laser-light examinations of the

FIGURE 3.5
GISP2 camp. Some days, life is difficult on the ice sheet.

ice, cutting ice slices for chemical and isotopic study, pulling off chunks of ice to analyze the gases they contain, slicing thin pieces of ice to look at their crystalline structure, measuring ice density and the speed of sound in the ice, and more. Many of the chemical analyses were also conducted at the site. Cores coming out of the drill hole required time to relax, as described in chapter 5, and to dry from the fluid placed in the hole to keep it from collapsing. Storage was needed for cores waiting to be processed, and for cores waiting to be shipped home after processing. The whole operation had to be protected from the wind and sun—the drifting snow of the first storm would have buried anything left on the surface, and the sun's warmth may have heated the ice closer to the melting point than we wanted.

The entire processing line was placed in a trench that had been cut twenty feet deep into the snow using a giant snow-blower, and then roofed with beams and boards. Snowfall slowly buried the trench, requiring several extensions for our entrance stairway and our cargo ramp/emergency exit. The roof of the trench was slowly squashed toward the floor by the weight of the snow piling up, but the design served admirably throughout the project and remained accessible for a few years afterward.

Cores were placed in special trays designed for the project, and the trays with ice were then pushed along on roller tracks, from saw to electrical-conductivity measurement, visual examination, and packaging, with slices shunted off into special side alcoves for further analyses. Electrical lights kept the trench bright. The temperature stayed at a "comfortable" twenty below. A CD player gave us something to talk about. ("Pink Floyd next?" "Anything *but* Pink Floyd!" "Pink Floyd it is." "Then I'm playing the Cow Tape next, with people mooing to music. . . .")

The side alcoves would make any dean or chief executive officer very happy. If the space wasn't big enough, we could pull out a saw (electrical chain saw or hand-operated carpenter's saw), cut some blocks out of the wall to make more room, and haul the blocks up to the surface.

Drilling ran day and night (Figure 3.6), with two crews splitting duties to push the hole ever deeper. Core drilled at night was stored until the day shift came on. For ten or more hours each day, the science team would push core down the line, analyzing, scrutinizing, and then packaging for shipping home. Evenings were for thinking about the day's results, repairing or improving equipment, copying data, and so forth. Work usually ran six days per week. Five summers—1989 through 1993—produced two miles of core (Figure 3.7).

Flights back to the states often carried ice on "cold decks"—no heat was turned on to warm the planes at the frigid altitudes where they fly. Special styrofoam-insulated boxes protected the ice during landing until it could be transferred to freezer trucks that met the planes and hauled the ice to cold storage, eventually at the National Ice Core Laboratory.

FIGURE 3.6
The midnight sun shines over the GISP2 drill tower in central Greenland.

FIGURE 3.7
Distant view of the drill rig shown in figure 3.8, showing the
vastness of the ice sheets and the relative insignificance of the
investigators.

Twenty miles to our east, the European GRIP team was
doing basically the same thing. Comparison between the cores actually proved to be critical—the two records are almost
identical for the most recent 110,000 years, but the records
differ for ice older than 110,000 years because ice flow has
mixed up the deepest layers in both cores. Two holes also
provided more ice, allowing more analyses and producing
more knowledge. The European GRIP effort had one big ad-
vantage over that of the United States—the Europeans had
their own GRIP label Italian wine (Giacomo Fenocchio &
Figli, Dolcetto d'Alba 1989). The U.S. camp was mostly dry,
and those desiring a libation had to bring their own from the
coast. Whether wet or dry, each camp succeeded in pulling a
two-mile stick of ice out of the ice sheet.

Drilling Deeper

When faced with a two-mile-high pile of ice, how do you pull out a 5.2-inch-diameter piece? People old enough to remember defrosting freezers appreciate the difficulty of chipping through even an inch or two of ice, let alone two miles!

Actually, ice-core drilling is not too difficult in principle. Take a metal pipe, cut some teeth on the end, hold the teeth against the ice and spin the pipe. The teeth cut chips of ice, and the core goes up inside the pipe. The idea is similar to that for coring through a door for doorknob installation; hardware stores sell short, toothy pipes that you can use as drill bits for this purpose. But in drilling ice, there are a few more problems than with doors (see Figure 3.8).

The weight of the ice will squeeze the hole closed unless the hole is filled with a fluid of nearly the same density as the

FIGURE 3.8
A lightweight drill used for taking ice cores to a few hundred feet deep. The larger drills for few-mile ice cores are similar in principle. This drill is being operated by Karl Kuivinen, Bill Boller, and John Litwak of the Polar Ice Coring Office, on ice stream B, West Antarctica.

ice. We used butyl acetate for this purpose. This is an environmentally friendly, rather nontoxic organic liquid that will not contaminate the ice for most studies, has low enough viscosity to allow a drill to drop rapidly through on the two-mile trip to the bed, and will not poison the drillers.

Just as sawdust must be removed to prevent it from clogging a wood drill, so must ice chips be flushed away from the drill cutters. Spirals on the outside of the rotating drill barrel, like the spirals on a drill bit for wood, flushed the chips up to a special holding chamber on the drill, aided by an industrial pump. Because the drill was suspended on a flexible cable, special leaf springs were required to push against the hole wall to prevent the motor and cable, rather than the drill cutters, from spinning.

Altogether, the entire drill string was nearly one hundred feet long, and hung from a hundred-foot spiderwork drill tower that dominated the GISP2 camp. The tower protruded seventy feet above a white geodesic dome that mostly kept the sun, wind, and drifting snow off the drillers. But on windy days, when the drill was pulled out of the hole, someone would have to climb above the dome, wrap a strap around the drill, and hang on so the drill wouldn't blow around and hurt people or itself. And that someone was a driller—see Figure 3.9.

Drillers are not exactly like other people. They tend to be smart, self-confident, experienced, talented—and different. Talk to a good driller for a while and you're almost certain to learn something that amazes you—he or she will start telling you about putting up some new route in mountain climbing, or building an airplane, or patenting a new invention, or sea-kayaking some untried route, and you will be thinking, "Wow!" The driller's job description includes building towers, handling tons of supplies, fine-tuning an instrument big and strong enough to kill someone, and producing core that eventually totals forty tons and is of high enough quality to keep the pickiest scientists happy.

The drillers' can-do attitude—"We're here, we have a

problem, we have some resources, we're a team, we will solve it"—extends right across the polar community. Whether cooking for fifty seriously hungry people, directing a cargo plane into the site, digging out the cargo plane after it sinks in soft snow off the side of the skiway, pulling five tons of cargo off the plane by hand to make it easier to get the plane going again, organizing the five tons of cargo before it gets buried by the snowdrifts from a big storm, fixing broken pieces of the plane right there in the snow, fixing nearly-antique heavy equipment, or inventing a Fourth of July barbecue-on-ice, polar drillers get the job done quickly, safely, efficiently, and usually cheerfully.

There are exceptions, of course, such as the fellow who became inebriated and relieved himself on a heater in one of the few warm buildings, but he caught a one-way trip to the coast on the first available plane. And such instances are in-

FIGURE 3.9
Catherine Melville, of the
Polar Ice Coring Office,
working on the GISP2 drill
inside the drill dome.

deed rare. On my returns from Greenland, my first encounter with a surly staffer at a fast-food joint on the Pennsylvania Turnpike was always an eye-opener, and I found myself wondering why the "real world" couldn't be a bit more like GISP2.

4

The ice of Greenland gives us incomparable records of past climates in and beyond Greenland. Tree-ring-like layers tell us how old the ice is. We can read how temperature and snowfall have changed in Greenland. Dust and sea salt in the ice were transported to Greenland by wind, and changes in their concentration in the ice tell us about changes in the winds that brought them. Bubbles of old air trapped in the ice contain information about past changes in the composition of the atmosphere itself. Recent layers are thick, allowing us to learn about individual seasons or even individual storms; older layers have been thinned by the flow of the ice sheet, so we can study longer times easily, although in less detail. In this chapter, we will consider how an ice sheet works. We'll see that gravity causes thick piles of ice to flow. This keeps Earth from becoming top-heavy and rolling over, and it also means that the deep ice of an ice sheet has a very long history.

One-tenth of the land on Earth today is buried by permanent ice. Ice grows wherever snowfall exceeds melting over many years. The snow piles up, is squeezed into ice under the weight of more snow, and eventually begins to flow under gravity. We call a flowing pile of snow and ice a glacier if it is small, or an ice sheet if it is really big. A little of the world's permanent ice is scattered around in various small glaciers in the Rockies, the Alps, the Andes, the Himalaya, and several other mountain ranges. But more than 99 percent of Earth's ice is in the great ice sheets that cover most of the island of Greenland and nearly all of the continent of Antarctica.

If all of the ice on Earth today were to melt, global sea level would rise about 200 feet vertically—not "Waterworld," but certainly a disaster, with the coast of Florida moving somewhere up into Georgia. Fortunately, we do not expect such a disaster over at least the next few centuries. There is a possibility that the West Antarctic ice sheet could collapse and raise sea level twenty feet in the next century or few centuries, but my reading of the recent literature suggests that this is not highly likely. Some sea-level rise in the future is almost unavoidable, however. The warming that occurred during the twentieth century has caused sea level to rise, as mountain glaciers melted and ocean waters warmed and expanded. More ocean warming and more glacier melting will occur even if the air temperature stabilizes—just as a winter snowfall takes a while to melt after the weather warms, the glaciers and oceans have not yet come into balance with the modern air temperature. Further warming of the air likely will bring even more sea-level rise.

The GISP2 ice coring project received some attention in the popular press during the early 1990s. After an appearance on radio or TV, we often received inquiries or suggestions from citizens concerned about the environment. Many of these were quite perceptive and interesting, but two common misconceptions deserve mention. The first is that we must be entirely wrong in claiming that the ice sheets are more than 110,000 years old: Airplanes that landed on the Greenland ice sheet during World War II have already been buried a couple of hundred feet, so the few thousand feet of the ice sheet can thus represent only a few centuries of snowfall. The second common misconception is that continuing snowfall on the ice sheets is causing them to build up, and pretty soon they will make Earth top-heavy and cause the planet to roll over. Back in the days of the Soviet empire, one group even sent out mailings arguing that the only way to save Earth from rolling over was to use U.S. and Soviet aircraft carriers to haul Antarctic icebergs to the Sahara.

There are several major reasons why these worries are misplaced, including the fact that the World War II planes landed in one of the regions of Greenland where snow accumulates fastest. But the main reason is that the well-meaning people behind these worries missed one cardinal fact about glaciers: Glaciers flow.

All Downhill

Consider, for a moment, pouring some pancake batter to make a mound in the middle of a cold griddle. As you pour, the pancake batter will begin to flow and spread. Gravity pulls on all piles and tends to flatten them out, causing the pancake batter to flow from where its surface is high to where its surface is low. Steeper or thicker piles tend to flatten out faster.

It may appear odd that ice can flow at all. We know that pancake batter is just a "thick" liquid, but ice is a solid—we can walk on it, drive snowmobiles on it, land huge planes on it, and generally treat it as a solid. The key to ice flowing is that it is a very warm solid—in fact, one of the hottest natural solids around.

"Hot ice" may seem strange to anyone who has ever sat on a snowdrift on an outhouse seat, but it is true. In discussing how materials behave, a "cold" solid is one that is far below its melting point, and a "hot" solid is one that is close to the melting point. In a freezer, an iron horseshoe and a chocolate bar are both stiff and brittle, and neither will flow. In your back pocket, though, the horseshoe will remain stiff and brittle, but the chocolate bar will "smoosh" as it warms near its melting point. Have a blacksmith heat the horseshoe white-hot, almost to melting, and the horseshoe will become nearly as soft as the pocketed chocolate bar. Because ice is typically within a few degrees or tens of degrees of melting, it is more like white-hot iron or a pocketed chocolate bar than a

shoe on a horse or a chocolate bar fresh from the freezer—
ice can flow. The flow of ice isn't fast, but it happens.

You could prove to yourself that the ice flows if you
placed some tall poles in the ice in various places on Greenland, and then used surveying tools to measure the locations
of the poles quite accurately over a few years. Many people
have done this experiment, and they find that all the poles,
and the snow and ice in which the poles sit, are moving away
from the center of Greenland toward the coast. Flow speeds
typically are a few feet or tens of feet per year, although some
poles will move as quickly as a few miles per year at Jakobshavn on the west coast of Greenland, which has the fastest average speed of any land ice on Earth.

About half the snow that falls on Greenland melts very
near the coast and feeds meltwater streams, and the other half
makes icebergs that float away and melt elsewhere. Flow
takes snow and ice from the high, cold center to the low,
warm edge of the ice sheet. Measurements show that the flow
very nearly balances the snowfall, so the ice sheet is not
building up and the world will not roll over.

The pattern of this flow also explains why most of the
history of the ice sheet is contained in layers deeper than the
World War II airplanes buried in the ice. To see the significance of this pattern, let's go back to pouring pancake batter
on a cold griddle. If you put a blob of batter in the middle of
the griddle, the initially steep, thick pile of batter will first
spread and thin rapidly, and then more slowly. A second blob
then placed on top of the first will also spread and thin, while
the first layer beneath it continues to spread and thin under
their combined weights. Adding more initially identical blobs
would make a pile of layers, with the oldest one on the bottom being the thinnest because it has been spreading the
longest.

A blueberry dropped into the pancake batter at some
time would be buried by later batter, moving downward toward the griddle surface as the batter beneath it thinned, and
moving toward the edge of the griddle and eventually falling
off the edge as the batter beneath it spread. The downward

motion of the berry would initially be rapid, because the whole pile beneath it would be spreading and thinning; later, as the berry approached the griddle, the downward motion would slow because the berry would have less batter thinning beneath it.

This is a fairly good approximation of the Greenland ice sheet. There is a continental divide running down the center of Greenland along the crest of the ice sheet, with snow and ice on the east side flowing east, and snow and ice on the west side flowing west; far to the north, ice flows north to the coast, and far to the south, ice flows south. I've walked, skied, and snowmobiled across the crest of Greenland, and I've never fallen into a crevasse because there are no crevasses near the ice-sheet crest. Yet east-side snow goes east, and west-side snow goes west. The layers of snow are stretched and thinned over time, just as stretching a rubber band thins it (as illustrated in Figure 4.1), or as a thick, steep blob of pancake batter flows into a wide, flat pancake. The bottom layer

FIGURE 4.1
Stretching a rubber band causes it to thin, as shown. If the bottom of the rubber band remains in contact with the table, then the top moves down. In the same way, ice moving west on the west side of Greenland and ice moving east on the east side of Greenland cause the layers between to stretch and thin, which in turn causes the top of the ice sheet to move down and make room for more snowfall.

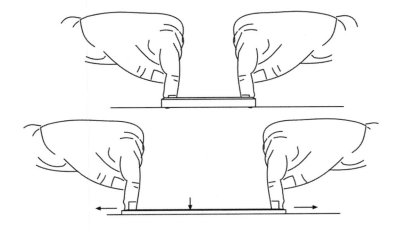

of the ice remains in touch with the rock beneath it, so thinning of the ice causes the upper surface to move downward.

The ice sheet has been transporting snowfall to the coast for a long time, and the shape and speed of the ice sheet have adjusted to move the snowfall it receives. The spreading and thinning of the ice move its upper surface down just enough each year to make room for the next year's snowfall. Any marker placed on the surface of the ice sheet, such as a blueberry or a drilling camp, will be buried deeper and deeper over time as the marker is transported toward the coast, but the shape of the ice sheet doesn't change much over time. Icebergs and surface melting remove ice near the coast, just as pancake batter runs off the side of your griddle.

Snow is compressed into ice under the weight of more snowfall in the top 200 feet or so of the ice sheet, over a century or two. A little bit of the air in the snow is "bottled" in bubbles to give us samples of old air, but most of the air in snow is squeezed out to the atmosphere. It is easiest for us to ignore the air here, and talk about the thickness of ice that is added each year. A three-foot snow layer that falls in a year in central Greenland is not quite as "airy" as some U.S. snowfalls, having about two feet of air and one foot of ice. This will be squeezed into a one-foot-thick ice layer in a century or two, so we can just say that one foot of ice accumulates each year.

Once that foot-thick ice layer has been buried halfway through the ice sheet, the layer has been stretched and thinned to half a foot in thickness; as the layer stretches, the ends melt very near the coast or break off to form icebergs that float away, so the ice sheet doesn't get wider. When the initially one-foot-thick layer is buried three-quarters of the way through the ice sheet, the layer has lost three-quarters of its thickness and is only one-quarter foot thick. When buried seven-eighths of the way through the ice sheet, the layer has lost seven-eighths of its original thickness and is only one-eighth foot thick, and so on. Thus, a core section from seven-eighths of the way through the ice sheet will hold eight times as many years as the same-length section from near the surface.

In reality, things aren't quite this simple. The thickness of a layer depends on how thick it was initially as well as how much it has thinned, as we will see. The model described here is a bit off in the oldest ice, especially if ice flow over irregular bedrock bumps has mixed ice of different ages—think of pancake batter flowing over a waffle iron rather than a smooth griddle.

But through most of the thickness of the ice sheet, each year's layer moves down just enough to make room for the next year's accumulation. The downward motion is caused by the stretching and thinning of all the layers beneath (Figure 4.2). Layers near the bed are very thin, stretch and thin only a little, and don't move down much; layers farther from the bed are thicker, and stretch, thin, and move down more. Any marker in the ice—a volcanic horizon or a World War II airplane—will be buried rapidly initially, but then more and more slowly as the marker gets closer and closer to the bed, just as the blueberry on your pancake batter was initially buried rapidly and then more slowly.

So if anyone ever warns you about the coming cataclysm as Earth rolls over, or tells you that buried planes show that the ice sheet on Greenland is young, you will know that flowing ice proves them wrong. Ice flows back to the sea about as rapidly as new ice is delivered by snowstorms, and the layer thinning caused by the outward flow of the ice from the center of the island means that the ice beneath the planes accumulated over vastly longer times than did the ice above the planes.

Piling It On

In the next chapter, we will discuss how it is possible to learn the age of ice cores by counting annual layers, much the way we can count rings in trees to learn how old the trees are. The thickness of an annual layer in a tree tells how much the tree grew that year. The thickness of an annual layer in an ice core tells how much snow accumulated that year, if we can "correct" for the thinning of the layer caused by ice flow, and for the amount of air the layer contains.

Correcting for air is really easy. Rather than dealing with thickness, we weigh samples and calculate how much ice is present, because air doesn't weigh enough to matter.

The details of flow corrections are rather mathematically and physically complex, and so are a bit beyond the scope of this book. The broad outline is given in the previous section—stretching progressively thins the layers. If an ice sheet

FIGURE 4.2

Cross-section through the Greenland ice sheet. The arrows show relative motion. Ice in the middle of the ice sheet, beneath GRIP, moves down. Ice west of GRIP, toward and beyond GISP2, moves down and horizontally to the west, while ice east of GRIP moves down and east. Layers are initially thick, as shown by the white band along the top of the ice sheet, but the layers thin as they are stretched and buried, as shown by the deeper, thinner white bands.

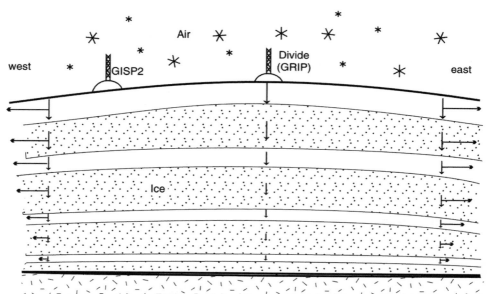

has not changed much since the snow in a layer fell, then the pattern of stretching and thinning also will not have changed much, and we can readily calculate how much a layer has been thinned. If large changes have occurred in the size or shape of the ice sheet, then the rates of thinning also will have changed over time, and we may have grave difficulties in figuring out the total thinning that has occurred.

Fortunately, central Greenland turns out to be a wonderful place to do such calculations. The ice sheet already almost completely covers the island. No one has ever observed an ice sheet that advanced far over deep ocean waters such as those around Greenland. Sediment cores from beneath those ocean waters show that they have not been overrun by the ice sheet, so we are confident that the ice sheet has not been much bigger in the past. Some evidence shows that the ice sheet was a little bigger during the colder times of ice ages and may have been smaller during warmer times further back, but there have not been any large changes over tens of thousands of years or longer.

Several different research groups have conducted calculations of how annual layers have thinned since their snow fell on Greenland. All agree quite closely. Thus, by understanding ice flow and dating the ice, we can get a wonderful year-by-year record of how much snow has accumulated on Greenland in the past. This is a useful indicator of precipitation, an important part of the climate, and it also is useful in understanding other climate indicators, as we will see soon.

5

History is the story of what events occurred, and of when those events occurred. In this chapter, we will discuss how climate historians date ice cores and other sediments, and in the next chapters we will go on to see how to read the story of what happened.

Many techniques have been developed for "dating" deposits, using changes in radioactivity or in chemical, physical, or biological characteristics that evolve at a known rate. The great number of dating techniques would form a book all their own, and we will not delve into them here. Suffice it to say that many different techniques have been tested against each other, against historical records, against common sense, and against annual records, and these dating techniques work well if they are applied carefully.

Some special sediments preserve annual layers, and the most direct dating technique involves counting these layers. We have now counted annual layers, and learned when things happened, with high accuracy for more than 100,000 years. Comparisons with historical events, and among ice-core, lake-sediment, ocean-sediment, and tree-ring records, show how reliable these dates are, as discussed next.

One, Two, Three, Four, Five, Six, Seven, . . .

Who among us has not counted the annual rings in a tree stump to learn how old the tree was when it was cut? Trees in seasonal climates are reliable calendars, marking each year with a new ring. Fortunately, the calendar can be read without cutting the tree. A pencil-width core can be safely removed from a living tree using a special coring device, and the growth rings in the core reveal how long the tree has been living. The tree may be growing on something interesting, such as a prehistoric grave or the pile of debris left by a melting glacier, in which case one learns that the feature of interest is older than the number of rings in the tree.

Occasionally, such a core from a tree will have a few missing rings, because an insect ate the wood, or a fire burned some of the wood away, or something else happened. But comparison of several cores from each of several nearby trees will quickly reveal where the problems are and allow dating with no errors.

In comparing samples from adjacent trees, one relies on a pattern of thick and thin rings. A tree grows rapidly and forms a thick layer in a good year, but adds only a thin layer in a bad year. A bad year may be too cold, too dry, or otherwise unfavorable. Nearby trees typically suffer or flourish together. Trees thus can be used for dating and for reconstructing past climates.

The best trees for recording climate are those that are stressed for some reason. For example, some trees live in places that are almost too cold for survival. For such trees, a "bad" year is a cold year, a "good" year is a warm year, and the history of tree-ring widths gives the history of temperature. Similarly, for a tree growing where the temperature is fine but water is scarce, tree-ring widths record rainfall.

The longest such records from living trees are only about 5,000 years long, from bristlecone pines in the mountains of the western United States. However, there is plenty of older wood around—in archaeological sites, buried in sedimentary deposits, or even sitting on the surface next to living trees.

The unique pattern of thick and thin rings from good and bad years can be matched between the overlapping parts of a living tree and a nearby dead one, allowing longer records. The longest such records now extend about 12,000 years.

Great, painstaking effort is needed to assemble such a long record. Confident interpretation requires that several cores from several trees be used for each time period, to make sure that insect or fire damage or other accidents don't introduce errors. With further work, it is possible that records longer than 12,000 years will be developed. Older wood with nonoverlapping tree-ring histories is available, but building continuous records isn't easy.

Trees are not the only calendars in the world—some sediments also are annually layered. In cold regions, streams wash sand into lakes during the summer but freeze during the winter, and the only sediments settling out of the lakes during winter are the smallest pieces that may take months to fall. Many lake sediments are stirred up by waves or burrowing animals, thus ruining annual records, but some lakes are not disturbed. In cold-region lakes with undisturbed sediments, each year is marked by a sandy summer layer and a finer-grained winter layer. Lake cores have been recovered with tens of thousands of annual layers. Occasional dating errors may occur, but rather accurate dating over long times is possible.

Colder Calendars

The longest continuous annual records yet recovered have come from our Greenland ice cores. Just as with tree rings and lake sediments, one can see annual layers in the ice.

How are these layers formed? In central Greenland, snow falls frequently throughout the year. The sun never sets in the summer, and never rises in the winter. Winter snow is buried without experiencing sunshine, but summer snow is "cooked" by the intense summer sun. This solar heating changes the snow, making visible layers.

The June noonday sun isn't strong enough to melt the

snow in central Greenland—that happens only about once every few centuries—but the summer sun does heat the snow within an inch of the surface as much as five degrees warmer than the air above or the winter snow a foot below. Much of the snow in the sun-warmed layer evaporates (also called sublimation), causing the layer to become light and airy.

Sublimation of ice is familiar to anyone who has left an ice-cube tray in a frost-free freezer for a few weeks. The ice cubes disappear as they form vapor, which moves through the air and grows frost on the colder, chilling unit of the freezer. This frost is removed occasionally by a quick melt-and-drip cycle. On the ice sheet, most of the vapor from the sun-warmed snow moves up into the air. When the sun drops low toward midnight, the snow surface and the air just above it cool as they radiate heat to space. Fog forms above the cooled surface, and frost grows from the fog. Frost forms on the snow surface (see Figure 5.1), and on the guy wires of the drill tower, the volleyball net, and everything else. The frost on the volleyball net later falls off, but frost on the snow surface usually is preserved when more snow falls on top.

The daily heat and nightly cold of the summertime thus turn an inch of wind-packed, fine-grained snow into two inches of coarse-grained, low-density snow called hoarfrost, or just hoar. Because the transformation from ordinary snow to hoar is driven by the sun, which shines only in the summer, we have strong reason to believe that summer and winter snow are quite different.

Crowding the Calendars

The next step is to see how this summer signal is changed as it is buried. We use snow pits and then ice cores to sample all the different stages as snow is transformed to ice and the ice layers are stretched and thinned, crowding the layers together but leaving us a record to read.

Snow pits are the easiest way to observe layers in the snow. Take a square-end shovel (and for deeper pits, a carpenter's saw or special big-toothed snow saw) and hack a

FIGURE 5.1

The snow surface in central Greenland. On clear nights when the
sun dips low, hoarfrost grows, much like a really spectacular
backyard frost in the northern United States. Our studies show that
such hoarfrost forms only during the summer, when the relatively
warm daytime air can hold enough water vapor that "night-time"
cooling will grow a thick layer. Identification of these hoarfrost
layers in ice cores allows dating.

hole in the ground. Most workers favor six-foot-deep pits, because it becomes difficult to throw snow out of a deeper pit,
although heroes have dug pits deeper than twenty feet.

To really see layers in the snow, we dig two pits, each a
six-foot cube, separated by a wall only a foot or so thick.
Together, two such pits require moving about six tons of
snow, so great care is recommended at the end to avoid kick-
ing a hole in the thin wall, ruining the experiment and requiring yet more digging. Then, using timbers and plywood, we
hammer together a roof over one pit, slide under the roof,
and let a last piece of plywood fall over the entrance hole
(see Figure 5.2). The sun shines into the open pit and through
the thin wall between the two, casting the layers in that wall
into brilliant relief.

I have stood in such snow pits with dozens of people—

FIGURE 5.2

To study snow layers, we dig two six-foot-cube snow pits separated by a thin wall, build a plywood roof over one, crawl into it through a "door" such as the one shown, and pull a last board over the entrance. Sunlight shining into the other pit and through the wall makes the layers very easy to see.

drillers, journalists, and others—and so far, every visitor has been impressed. The snow is blue, something like the blue seen by deep-sea divers, an indescribable, almost achingly beautiful blue. Water, whether liquid or ice, absorbs red light a bit more than blue. Let light penetrate tens of feet into the ocean and the red is filtered out, so only blue reaches your eyes. In snow, a ray of light passes through a tiny crystal, is bent, goes through another, bounces off another, and so staggers on its way to the viewer, going much farther than the straight-line distance to the eye. On the way, the red is lost, and the result is a beautiful blue.

After the blue, the next thing most people notice is the layering. The low-density, coarse-grained hoar layers formed by the sun's heat show up as light bands across the pit. High-density, fine-grained snow packed by the winds of storms appears darker. Even subtle differences in grain size or density cause slight differences in the way the snow scatters the light, and thus in the appearance. Where some former windstorm has scoured the top of a hoar layer, the dark snow occupies a scoop out of the light. Where a snowdrift built across the surface, faint slanting beds are visible within a dark layer, as seen in Figure 5.3. One year, the late-summer crew in camp became a bit careless about staying out of the designated "clean-air zone—skis and foot travel only" during some unauthorized recreational activities involving snowmobiles. When we happened to dig through one of the unapproved snowmobile tracks the next summer, the slight crushing of the hoar layers in this modern "dinosaur track" was clearly visible.

In central Greenland snow pits, a foot or two of nearly homogeneous winter snow is followed by several inches of summer snow from a few snowfalls, with a coarse, airy hoar layer at the top of each summer snowfall. Two to three years of snow fit in a six-foot snow pit. In other places, a six-foot-

FIGURE 5.3

Here is the wall of the snow pit in the previous picture. The lighter-colored layers are summertime hoarfrost, and the darker layers are snow packed by the wind. The pit was dug early in the summer of 1992, and is about six feet deep. The light layers about two feet down mark the summer of 1991, and the light layers about four feet down mark the summer of 1990.

tall snow pit wall may reveal less than a year or more than a decade, depending on how much snow falls there each year.

We have several ways to test the reliability of these results. The high-tech ones will be described below, but there are several simple ones. For example, a bamboo pole stuck into the snow surface will be partially buried the next year. Measure how much of the pole sticks out one year, and how much remains sticking out the next year, and you know how much it snowed. This thickness of new snow is found to equal the distance between one cluster of hoar layers and the next in a snow pit. We can also look for summertime indicators such as snowmobile tracks from previous years. The first winter-over expedition to the GISP2 site did not occur until a few years after our snow-pit digging, so the snowmobile tracks we saw could have been formed only in the summer.

The distinctive appearance of summer snow, caused by the summer sun, is preserved, buried, and easily recognized in snow pits. But are these differences recognizable in an ice core? A two- to three-foot-thick layer of freshly fallen central-Greenland snow is squeezed to a foot or a bit less in thickness as it is changed to ice. The crystals in it grow, the interconnected air spaces are pinched off to form individual bubbles, and other changes occur. Almost surprisingly, though, the annual layers remain visible.

The easiest way to look at an ice core is to place it on a light table—a sheet of glass or clear plastic over a light source, such as fluorescent bulbs, as shown in Figure 5.4. We sometimes added fiber-optic illuminators—high-tech flashlights—to highlight those places not seen clearly in the fluorescent light.

When examined in this way, cores from very shallow depths look just like snow in a snow pit, but without the wonderful blue color, as the 5-inch-diameter core is not thick enough to filter out much of the red light. Summer snow is marked by many light, coarse-grained layers, and winter snow is nearly homogeneous and appears darker. As one goes deeper, the compacted snow is called *firn* (a German word for old snow). The weight of the overlying snow has squeezed the ice grains together, but air can still move through the spaces between the grains. Eventually, after almost 200 years when the snow has been buried about 200 feet down, the firn is squeezed enough to seal off and trap the remaining air as bubbles, and we say that the firn has become ice.

As the snow changes to firn and then ice down the core, summer and winter layers remain distinct. The coarse-grained, low-density snow layers of summer produce big-bubbled layers of ice. Just as by counting tree rings, one can tally the summers of history.

About two thousand feet down into the ice of central Greenland, things become slightly more difficult. The weight on this 2,000-year-old ice is more than half a ton per square inch, and the air in the bubbles has been compressed so that it pushes back with this pressure. Cores brought to the surface will pop, snap, and break as the air expands and fractures the

FIGURE 5.4
Kurt Cuffey, then of Penn State and now a professor at the University of California–Berkeley, inspects an ice core on a light table in the undersnow laboratory at GISP2. Careful examination by Cuffey and many others revealed annual layers that allowed accurate dating of the ice core.

ice. (Many old-time polar hands have obtained a little such ice, from an iceberg or elsewhere, and enjoyed the naturally fizzy, noisy drink that results when alcoholic beverages rapidly dissolve the bubble walls and the air "pops" out.) There is little to do with such brittle ice except set it aside and let it relax as the bubbles expand and their pressure drops for a year or so, during which time the ice often breaks into smaller chunks. We then practice jigsaw-puzzle assembly and put the cores back together again. Amazingly, when the jigsaw-puzzle ice is examined on the light table, the summer layers are still there, staring through the cracks.

At just less than a mile deep in central Greenland, in 8,000-year-old layers, the pressure becomes so great that the air in the bubbles begins to dissolve into the ice. In a bit of the ice right next to a bubble, the hexagonal "snowflake" arrangement of water molecules changes to a roomier cubic structure, and the air molecules slip into the spaces in the centers of the cubes. The bubbles disappear, replaced by bubble-sized pieces of an ice-air mixture called *gas-hydrate* or *clathrate*. The clathrate, which looks and acts almost like ice, forms because the air takes up less space in the clathrate than it does in a bubble. As clathrates replace bubbles, the ice cores become as clear and easy to handle as plexiglass tubes. But leave this ice on the surface for a few months, and the bubbles begin to reappear as the clathrates break down.

The observer actually has two choices for counting annual layers in mile-deep ice from Greenland. The obvious one is to wait for the bubbles to come back and then find the big-bubbled summer layers. But careful examination of the core as soon as it comes out of the drill shows something else. Faint, grayish, ghostly bands are visible in the focused beam of an intense fiber-optic lamp. These are the late-winter dusty layers, rich in soil particles blown onto the ice sheet from fields and deserts of Asia and elsewhere. As one goes deeper, the ice becomes dustier, and these bands are easier to see. In ice from the cold, dry, windy ice age, the bands are so strong that they can be counted from across the room, without any special lighting. As the bubbles return, the gray bands become harder to see, although they are still there.

Charging Ahead

Dating a two-mile-long ice core is an involved task. Simply looking at two miles of ice, a few feet at a time, takes many months and involves a high level of tedium. For the GISP2 ice core, the overall dating effort involved more than a dozen people, a number of techniques, several years, and a few friendly arguments. We started using four different dating methods: visible layers, identification of volcanic fallout, electrical conductivity, and ice-isotopic ratios (a thermometer we will discuss soon).

Appearance is one of a host of differences between summer and winter snow. Dust content, chemical composition, and isotopic ratios all change with the seasons. For example, during the winter when a thin layer of ice grown from sea water floats on the ocean around Greenland, not much sea salt blows onto the ice sheet. As the ocean surface begins thawing in late winter to spring, strong storms supply salt to the ice sheet.

Hydrogen peroxide is produced in the air by chemical reactions driven by sunlight and promptly falls on the ice sheet, so peroxide occurs in summer snow but not in snow from the sunless winter. The ice-isotopic thermometer shows the clear difference between winter and summer temperatures.

One "easy" way to measure ice chemistry in the search for annual wiggles is through the electrical conductivity of the ice. Most snow and ice are naturally weak acids ("acid rain" and "acid snow" involve a human-caused increase in the natural acidity). The air contains many chemicals, including carbon dioxide and sulfur dioxide. Carbon dioxide dissolves in rain or snow, producing a weak acid called carbonic acid, which is the acid primarily responsible for dissolving some rocks to make caves. The air also contains sulfur dioxide, put there by explosions of volcanoes and by other processes (recently, with a lot of help from humans). The sulfur dioxide reacts with water in the air, producing droplets of sulfuric acid, which may collide with snowflakes or raindrops, and even may serve as focal points on which snowflakes or raindrops grow.

The degree of acidity in snowfall changes from summer to winter as the chemistry of the atmosphere changes, so ice formed from accumulated snow also has varying levels of acidity from summer to winter. If you try to pass electricity through ice, the current never flows very easily, but the more acid the ice contains, the more easily the electricity can flow, in the same way that battery acid carries electricity more easily than ordinary water does. Thus, by measuring how easily electricity passes through the ice, one can measure the variations in acidity caused by the changing seasons.

Kendrick Taylor, of the Desert Research Institute in Reno, Nevada, made this measurement for GISP2, using an apparatus that is simple in principle but required careful engineering. He placed two electrodes against the upper surface of the ice with a high voltage between them, dragged them the length of a core section, and measured the current twenty-five times per inch as the electrodes moved. You might think of a nine-volt battery, one of the small, blocky ones often used in home smoke detectors. Each battery has two electrodes on top. Touch a battery to a tester, and the strength of the current that flows from one electrode to the other through the tester shows whether the battery is "good" or not. An older, no-longer-recommended way of testing such a battery was to turn it upside down and touch the electrodes to your tongue, which would immediately and unpleasantly feel the electricity if the battery was good. Replace the tongue with an ice core, raise the voltage, and you have a vague idea of Kendrick's apparatus. He could measure six feet of core in a minute or so, and show the record of the layering as a wiggly green line on his computer screen.

Kendrick's electrical conductivity method, or ECM, detects annual oscillations in the chemistry. It also finds highly conductive layers full of acids from major volcanic eruptions, and layers that do not conduct electricity well because their snow collected windblown dust or forest-fire smoke that neutralized the acids. An experienced worker can tell the annual wiggles from the occasional signal of forest fire or volcano, and thus can count years.

Every piece of core that rolled down the processing line stopped at Kendrick's ECM station for a few minutes, and he turned it into a nice green line and a piece of computer memory, as illustrated in Figure 5.5. For a while, Kendrick turned on an audio speaker and had the computer assign high notes to highly conductive layers and low notes to less conductive ones, so that we could listen to the music of history. Each evening, Kendrick and his assistants printed out the computer files and studied them.

FIGURE 5.5

A sketch of the apparatus used to measure electrical conductivity of ice cores. The core is cut along its length using a band saw, and the top piece removed for other analyses. Then, wire electrodes are dragged along the length of the cut on the remaining piece, and the current flowing between the electrodes is measured, recorded, and displayed on a computer.

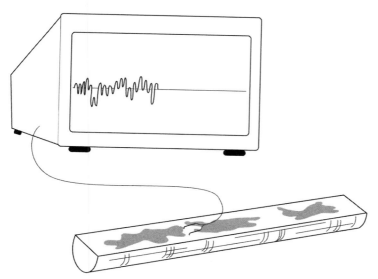

Seeing More Clearly

The cores were then rolled to the light table. There, I (and in later years, Tony Gow, Deb Meese, and others) marked down everything visible on three-foot-long strips of graph paper bound in books, making a full-size map of every core. And each time a summer layer went by, I marked it on the book and also made a tally on a separate sheet. We thus could keep track of the passing of history: This snow fell the year I was born, that snow when Lincoln spoke at Gettysburg, and so on.

Our first target was 1783, the year of the great fissure eruption of the volcano Laki in Iceland. From June 1783 through February 1784, a chasm in Iceland poured out cubic miles of lava flows and volcanic ash. Fire fountains arced hundreds of feet in the air, and a fine layer of ash covered the island and was recognizable one thousand miles to the east. Crop loss in Iceland caused widespread starvation. Unusual "dry" fogs containing less water than normal fogs were reported across Europe and into Africa, Asia, and North America. The fogs contained much sulfur dioxide, a common gas from Icelandic volcanoes, and snow that fell through the fogs collected sulfuric acid.

Early ice-core studies had shown that Laki dumped more sulfuric acid on Greenland than any other event that occurred within several centuries of Laki, producing a spike in electrical and chemical records that stands far above surrounding peaks. The Laki peak is a time marker that allows all cores in Greenland to be correlated, and allows dating to be checked. The date usually is taken to be 1783, although there is some disagreement on whether the acid remained in the air until 1784, so dating can be checked to within one year.

When the ECM electrodes scratched across 236.1 feet depth in the core, the current was rising, and it rose higher than any previous peak before falling back. Later work by volcanologist Greg Zielinski of the University of New Hampshire would show that minute glass fragments in the ice core had the same chemical composition as fragments known to be from Laki. Chemical studies on the ice would confirm that the electrodes were measuring the effects of a peak in sulfuric acid. But we already knew that this was Laki. My running tally based only on my observations of the core had us in the year 1788, an error of four or five years out of more than two hundred. That night, I pored over the maps of the cores and found that in the bustle of core processing, I had mapped three summers that I forgot to transfer to my tally sheet, so the error dropped to one or two years out of two hundred.

Months of work remained. Thousands of samples were being cut up, to be analyzed later for their ice-isotopic composition to reveal seasonal changes. Previous workers had shown that these isotopic ratios are especially useful in dating Greenland ice cores. Once the isotopic data became available, we compared counts based on ECM, isotopes, and visible appearance. We also compared these to the volcanic time markers, including fallout from the eruptions of other Icelandic volcanoes, Hekla, Katla, and Eldgja, and from the year 79 eruption of Vesuvius, the 1479 eruption of Mount St. Helens, and so on. Other data sets were added, and a dating committee pored over all the records. Occasional disagreements required adjustments and fine-tuning of the time scale. But standing in the trench watching the ECM roll up the peak of Laki, we knew we had joined the select ranks of those who can date paleoclimatic records accurately.

Seeing Further

There are no events older than about 2,000 years that are both reliably dated by historical records and that left a clear signal in the ice cores. We then must look for other techniques to verify our dating.

One is to make "internal" comparisons. These include having several people count layers several times using several different seasonal indicators to see how well they agree, without "cheating" by finding out what other people measured. These comparisons, both within our group and between our group and the GRIP collaboration, showed errors of only about one year in one hundred over the millennia of the recent warm period, with errors slowly increasing in older and deeper ice.

The counting errors probably increased in deeper, older ice for several reasons. First, as described earlier, the flow of the ice causes the layers to become thinner and thinner with increasing depth, which increases the likelihood that we will not see a layer. The flow of the deeper ice also has "messed up" certain layers, bending some into Z-shaped folds (Figure 5.6) and causing others to stretch and thin to the point where they are no longer continuous, thereby complicating inter-

FIGURE 5.6

Flow can disturb layers deep in the ice sheet, as shown in this picture of a section of the GISP2 ice core from about 1.5 miles deep. The lighter-colored bands are dust-rich layers, and the darker-colored bands are less dusty ice. The picture is about five inches high. Layers very near the ice sheet surface are bumpy because of snow drifting, but the stretching and thinning of the layers reduces those bumps so the layers are nearly flat and horizontal through most of the top mile of the ice sheet. Then, little wiggles begin to develop from ice flow, and to fold as shown here. Going deeper, some of the wiggles become larger, and eventually disrupt the continuous record of past climates. Fortunately, the disturbances give many clues that allow easy recognition, so our interpretations are not compromised. The white spot is a light.

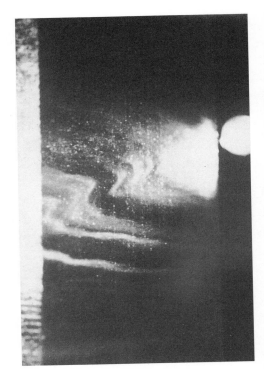

pretation. In addition, colder times in the past had less snowfall per year but more wind than recently, so it is possible that the accumulation from entire years blew away. Today, in the cold central areas of Antarctica where snow piles up especially slowly, annual layers are not preserved reliably. And despite heroic efforts, the quality of the ice cores was not always perfect, so measurements were difficult in some places. Even so, the various methods we tried for counting layers agreed with each other within a few years in one hundred over the most recent 50,000 years, and we could count well over 100,000 years, although with errors increasing, probably to as much as ten years in one hundred in the oldest ice.

Agreement does not prove accuracy, however; perhaps all of us were being fooled in the same way. To check for this possibility, we needed to look for other help. As we will see in subsequent chapters, the ice in Greenland has collected climatic information on many things: the temperature and snowfall rates in Greenland, how strong the winds were that carried sea salt from the ocean and dust from the deserts of central Asia, the wetness of subtropical regions where swamp gas (methane) was produced, and more. The ice core shows that at certain times, large, rapid changes occurred almost simultaneously in all these climatic indicators. One such time of rapid change occurred about 11,500 years ago, when Greenland ice cores record the sudden warming, increases in snowfall and methane, and drop in windblown materials that led from the Younger Dryas cold event to the modern climate. A near-simultaneous warming, wetting, and wind drop in many regions beyond Greenland is required to explain the changes we observed in the ice.

These changes also have been recognized in sediments from other regions. The warming caused glaciers to melt, exposing rocks beneath them and allowing trees to grow or lakes to form in once-glaciated regions. Beyond the reach of glaciers, the pollen and seeds falling into lake sediments changed from cold-weather types to warm-weather types as the temperature rose. Increased rainfall caused lakes to form in regions that once were dry, and to begin filling with sediments. Furthermore, these changes have been dated as occurring at the same time as the warming recorded in the ice core, within the dating uncertainties of a century or two, using layer counting and other dating techniques. Because our layer counts for this climate change agree with many other independent estimates of its age within our combined uncertainties, we have greater confidence in our results. Such agreement is found for ages of other climate changes as well as for the change 11,500 years ago, increasing our confidence.

In short, we can learn when changes occurred that left signals in the ice cores. And we can learn whether those changes occurred over centuries, decades, or merely years.

6

Temperature dominates much discussion of climate. Humans live differently in hot summers than in cold winters, and differently in hot tropics than in cold polar regions. Our ancestors certainly noticed the cold of the ice age, and we spend much effort debating whether we will be affected by human-caused greenhouse warming.

To study past climates and predict future ones, we wish to know temperatures in many places at many times. Patterns of temperature change will help reveal what caused climate shifts. For example, warming of most or all of the globe may result from an increase in greenhouse gases, but tropical cooling may accompany polar warming if a more vigorous atmospheric or oceanic circulation moves heat more efficiently from the tropics to the poles. Dust and swamp-gas methane trapped in ice cores tell us about winds and wetlands far beyond the ice sheets, but ice cores only record temperatures on the ice sheets.

Fortunately, there are many ways to estimate temperatures of the past on and beyond the ice sheets. Here, we will focus especially on the use of stable isotopes and borehole temperatures of ice to learn past temperatures on ice sheets, after a brief tour of some other techniques used elsewhere.

Written records are useful in reconstructing recent past climates. The canals of Holland were built for transportation, and official records have been kept of their freezing and thawing.

The descendants of Hans Brinker cannot do as much ice skating on the canals as he did, because the canals are not frozen as much now. Historical records, such as diaries and letters, show generally cold winters and widespread sea ice (frozen ocean water) in and near Iceland at the same time Hans Brinker was skating, in the Little Ice Age (roughly 100 to 500 years ago).

Before the existence of written records, we must turn to other indicators. Sediment in lakes and bogs contains pollen, twigs, leaves, and other remains of organisms that lived in and near the water. Because some plant types prefer cool conditions and others warm, some wet and some dry, the types of pollen, twigs, and seeds can be used to estimate past climates. Care must be taken to separate the effects of changes in temperature from changes in moisture availability, and to recognize that one living thing may disappear because some other type pushed it out (say, because humans arrived with fire and plow).

Numerous other techniques of paleothermometry exist, which are described in good introductory textbooks. In the oceans, for example, the types of creatures living and leaving their shells in sediment change as temperature changes. Also, some plants in the oceans build their cell walls of stiff molecules during warm times when the heat tends to soften the walls, but these plants use "floppy" molecules during cold times when the cold would make stiff walls brittle. These molecules survive being eaten and excreted, and eventually end up in sediments. The stiff:floppy ratio of these particular molecules in sediments thus is a paleothermometer.

Interestingly, the best paleothermometers are probably those on the ice sheets. However, to read that record, we need to take a brief detour into the wonderful world of stable isotopes.

Heavy Water

Almost everything around us is made of atoms, each of which contains a dense nucleus of heavy, positively charged protons and uncharged neutrons surrounded by a cloud of lightweight, negatively charged electrons. The number of electrons and protons balances, so that an atom is electrically neutral. Stealing or sharing of electrons causes atoms to stick together or be pushed apart, and is called *chemistry*.

As far as chemistry is concerned, all atoms of a type, or element, are nearly the same. The type is fixed by the number of protons in the nucleus: 1 for hydrogen, 8 for oxygen, and so on. Changing the number of protons creates a new element with different chemical behavior.

The neutrons in the nucleus keep the positively charged protons from repelling each other right out of the atom. Too few or too many neutrons will create an unstable, radioactive nucleus that eventually falls apart in some way, but the exact number of neutrons is usually not critical—some atoms of an element may have an extra neutron or two compared to others. Atoms of an element with different numbers of neutrons are called *isotopes*, from the Greek for "same place" because they belong in the same place in the chemically based periodic table of the elements. Oxygen, for example, has 8 protons, and either 8, 9, or 10 neutrons, for a total of 16, 17, or 18 heavy particles in the nuclei of oxygen-16, oxygen-17, and oxygen-18, respectively. Hydrogen has one proton, and comes in the no-neutron type hydrogen-1 and the one-neutron type hydrogen-2. For historical reasons, hydrogen-2 is also called *deuterium*, with "deuter" meaning two. There is even hydrogen-3, or *tritium*, but this isotope is unstable and decays radioactively over a few years to decades.

In combining two hydrogen atoms (mass 1 or 2) and one of oxygen (mass 16, 17, or 18) to make water, we can end up with water with an atomic weight of anywhere from 18 to 22. Because 99.8 percent of the oxygen is oxygen-16 and more than 99.9 percent of the hydrogen is hydrogen-1, most water weighs 18 and "heavy" water is rare, but all natural water samples include a little heavy water.

The physical behavior of a water molecule is affected slightly by its weight. The heavier a water molecule is, the slower that molecule moves and the harder time it has jumping out of liquid water to evaporate. In addition, the heavier a water molecule is, the more likely that molecule is to condense from vapor to make a raindrop or snowflake. Hence, water vapor is isotopically "lighter" than water or ice with which it is in equilibrium.

Isotopic compositions are measured using a mass spectrometer. For oxygen, for example, a sample of water or some other material containing oxygen is chemically converted to carbon dioxide in some way, because the carbon dioxide is easier to measure. An electron is then knocked off each carbon dioxide molecule by bombarding the molecules with more electrons, giving each carbon dioxide an electric charge. The charged carbon dioxide ions are pushed down a tube by an electric field, in much the same way as the picture tube in a television throws particles at the screen. On the way, the charged carbon dioxide is passed through a magnet, which pulls on the electric charges to bend the flight paths. A heavier carbon dioxide is harder to turn aside, so the carbon dioxide ions with heavy oxygen leave the magnet in a slightly different direction than do the ions with only light oxygen. The ions are then collected in "cups," and the electrical current supplied by the arriving ions is measured, vaguely in the same way as the brightness of a spot on a TV screen tells how many particles are hitting the screen there. The cups are placed to collect the heavy and the light carbon dioxide ions, and the ratio of heavy to light is calculated.

Long experience has shown that the easiest way to obtain good data, and to talk about those data, is by comparing the measurements of isotopic ratios to measurements of isotopic standards of known composition. It is easy for you to tell which spot on a TV is brightest, but difficult to describe how bright a spot is. A standard just gives you a spot for comparison. The standard for oxygen and hydrogen is a human-made approximation of the water in the ocean, called *standard mean ocean water*, and we ask whether a water sample is "heavier" or "lighter" than this standard.

A Light Frost

Stable isotopes are used to trace an amazing variety of physical processes, some of which we will discuss through this book. Perhaps the best known use of stable isotopes is in paleothermometry.

Earth's climate system is a way to move heat and moisture from the sun-drenched tropics to the cold polar regions. Air masses moving from the tropics to the poles contain moisture. These air masses cool on their way, by radiating heat to space or by warming the land or water beneath them. Cooling typically reduces the moisture content of air by causing growth of clouds that snow or rain on their way.

Isotopically, the water vapor in an air mass starts out a little lighter than that in the ocean, because isotopically lighter water has an easier time evaporating. The difference in ease of evaporation is small, however, so vapor does contain some heavy water. When the first raindrops fall from the first cloud, they average somewhat isotopically heavier than the water left in the air mass because heavier water condenses more easily. In fact, the first condensation simply reverses the initial evaporation, so the first rain has the same isotopic composition as the ocean water from which it came.

As rain and then snow are removed from an air mass, the precipitation is always isotopically heavier than the vapor it leaves behind. Thus, the remaining vapor becomes lighter and lighter as it becomes colder and colder, and the precipitation also becomes lighter as the air runs out of heavy isotopes to lose to precipitation.

The great isotopic geochemist Willi Dansgaard analyzed the yearly average isotopic composition of precipitation at a great number of sites around the world with known temperatures. In 1964, he published the results; if you gave him the average isotopic composition of precipitation where you live, he could tell you the average temperature where you live quite accurately. If he plotted the isotopic composition of precipitation against temperature, the points fell very close to a straight line.

This clearly opens the possibility of paleothermometry. If we can measure the isotopic composition of some water that fell as rain or snow sometime in the past, we have a thermometer for the time when the water fell. Old water can be found in the ice of an ice sheet, or in well water, or in water that was taken up by a tree as it grew and was used to make wood. Shell-building animals take oxygen from water and store it in their shells, so old shells record the composition of old water. This is a powerful technique, and is used routinely to learn past temperatures.

Any temperature estimate, for today or in the past, includes some uncertainty. If you have a thermometer hanging outside your window, you know that the temperature you read is not exactly correct. The dial may be rotated slightly or may be hard to read, the thermometer may register too warm when it is in the sun, and so forth. All data have such calibration errors, measurement errors, and biases. Technical reports typically include a statement of uncertainty whenever measurements are given.

In the same way, a paleoclimatologist looks for a "best" estimate of past temperature, and for an estimate of uncertainty. With the isotopic thermometer, there are many sources of uncertainty. Measurement errors are quite small, but not zero. More important, the isotopic composition of precipitation can be affected by more than just temperature. For example, summertime snow on ice sheets is isotopically heavier than wintertime snow because summer is warmer than winter. A switch from mostly summer snowfall to mostly winter snowfall would look like a cooling event even if the temperature didn't change. Modern observations and various models show that temperature is still the most important control on the isotopic thermometer, but that many things, including the seasonality of precipitation, can affect isotopic ratios. We thus must at least consider the possibility that the relation between temperature and isotopic ratio—the calibration of the thermometer—has changed over time.

Fortunately, on the ice sheets and in some other places, we have another thermometer. This is a more faithful thermometer than isotopes, because it is not affected by many extraneous factors. But it is a "fuzzy" thermometer—it "forgets" old temperature changes if they didn't last very long, whereas the isotopic ratios have a much better memory for short-lived changes. By combining these two thermometers, we obtain a better measure of past temperature.

Suppose that Grandma calls and says she would like to come over for dinner in an hour and she is really hungry. You rush to the freezer, pull out the roast you've been saving, and throw it in the oven. Half an hour later, the outside of the roast is sizzling and popping, but the inside is still cold, when Grandma calls back and says that she forgot about her jogging date with the nice man downstairs, and she won't be coming after all. So you grab the roast, toss it back in the freezer, and run off to chaperone Grandma's jogging date.

Ten minutes later, suppose that your spouse or significant other came home and somehow knew to drill a hole in the roast and measure the temperature at different depths along the hole. The center of the roast would still be cold from being in the freezer to start with, a zone around the center would be warm from the oven, but the roast would be cold right at the surface from its return to the freezer. Over time, the roast would slowly cool and freeze completely again. Until then, your spouse could do a pretty good job of figuring out the history of the surface temperature of the roast (freezer-oven-freezer), as seen in Figure 6.1. Your spouse would need to know something about freezers and roasts (such as the facts that the freezer is not beaming microwaves into the roast, and that the roast is not highly radioactive), but the rest is pretty easy.

Notice that if Grandma called you thirty times in twenty minutes and made you move the roast from oven to freezer or back each time, you would be a frustrated person and your spouse would not be able to figure out the whole story. Because heat spreads out, or diffuses, rather quickly, the records of those rapid changes would run together after a while; your spouse could tell that the roast had been in and out of the oven, but not how often or exactly when.

Now, suppose you were to go to the Greenland ice sheet, drill a two-mile-deep hole through it, wait a couple of years for the small amount of heat from your drilling to dissipate, and then drop a thermometer down the hole, measuring the

FIGURE 6.1

The history of surface temperature around a roast is shown in the left panel, as the roast is moved from freezer to oven and back to the freezer. If the temperature in the roast is then measured, the center of the roast will never have warmed up, the outside of the roast will already have cooled off, but a zone between the center and the outside will remain warm for a while, as shown by the cartoon in the right panel.

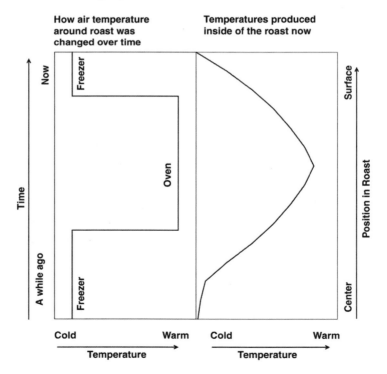

temperature at different depths. As you go down in the hole, you would expect the temperature to increase because of the heat flowing out of the earth beneath, in the same way that the deep parts of mines and oil wells are hot. Indeed, two miles down at the bottom, the ice is hotter than at the surface and only a few degrees below the freezing point. But one mile down, the ice is actually colder than at the surface. The cold zone a mile down is left from when the ice sheet was in the "freezer" of the most recent great ice age. This cold ice has been slowly warming over the 20,000 years since the coldest times, but still has not warmed to modern temperatures.

Going down in the ice sheet, the warmer-colder-warmer wiggles in temperature record the cold of the previous winter, the warmth of the twentieth century, the cold of the Little Ice Age, the warmth of a few thousand years ago, the cold of the last great ice age, and the warmth from before that ice age. The older one gets, the more the information is "smeared out," and the bigger and longer an event had to be to be recognizable, but the information is there (see Figure 6.2).

The temperature profile also depends on ice motion. As the ice deforms, friction in it makes a little heat. The moving ice tends to carry its temperature with it, in the same way that the roast carried its temperature from freezer to oven. But these are well-understood processes, and the corrections for them are rather straightforward. If the size and shape of the ice sheet had changed greatly in the past, the calculation would be much harder, but we have good reasons to believe that Greenland's ice has not changed greatly for the last 100,000 years, as we discussed in chapter 4.

There are many ways to "read" the temperature profile. All create difficulties and all suffer from nonuniqueness; the size and timing of the changes cannot be learned exactly, and the record of the old, short-lived changes is lost.

Fortunately, however, we now have two paleothermometers—the isotopes of the ice, and the temperature of the borehole. The common signal in these two records is the temperature history of central Greenland. The glaciologist Kurt

Cuffey, working first in my lab, then at the University of Washington and now at Berkeley, pioneered this technique. An independent European team achieved almost identical results. Kurt assumed that he understood heat flow in ice, and

FIGURE 6.2

The temperature history, and modern borehole temperature, for central Greenland. This situation is similar to that of the roast shown in the previous figure. A 50,000-year history of surface temperature inferred from the isotopes of the ice core is shown on the left, with today on the top. The temperatures measured today in the borehole from which the core was taken are shown on the right, from just over halfway through the ice sheet (bottom of the figure) to the surface (top). The light lines between the left and right panels show how the Little Ice Age, the warmth from before the Little Ice Age (including the Medieval Warm Period and earlier warm times, here labeled "Mid-Holocene warmth"), and the ice age are recorded in the ice isotopes and in the borehole temperatures. The warmth at the bottom of the borehole includes warmth from before the ice age, and warmth flowing up from the heat deep inside Earth.

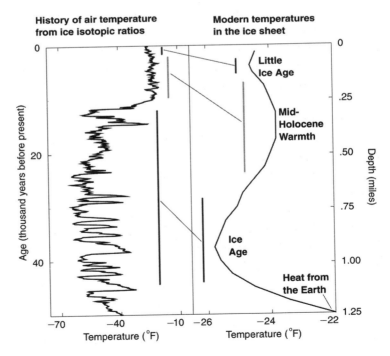

that the isotopic ratios are indeed records of temperature. He then made an educated guess of how the isotopic ratios were related to temperature, with every intention of coming back and correcting the guess.

With such an educated guess, the record of isotopic compositions from an ice core becomes a history of surface temperatures of the ice sheet. Just as the freezer-oven-freezer history of surface temperatures of the roast produced the cold-warm-cold temperatures in the roast when your spouse checked on it, any history of surface temperatures predicts the temperatures you will observe today at different depths in the ice sheet. The educated guess is used to calculate the temperatures it predicts in the ice sheet today, and these are compared to the temperatures actually measured in the borehole. The educated guess is then adjusted until the ice isotopes do the best possible job of predicting the borehole temperatures. These adjustments can be made using a computer-based approach called an inverse technique, and come rapidly to the best answer.

Thousands of measured ice-isotopic values are used to predict thousands of measured borehole temperatures, and only three numbers are adjusted—how much of an isotopic change occurs for a one-degree temperature change, the temperature at which precipitation would match the isotopic composition of ocean water, and how much heat is supplied to the bottom of the ice from deep inside Earth. If the isotopic ratios are able to predict the borehole temperatures accurately, then the isotopic ratios record the surface temperature when they fell and not other things—even if the three "knobs" are twiddled, they could not make thousands of numbers match thousands of other numbers by accident. If the isotopic ratios fail to predict the borehole profile, then the isotopes are not a good thermometer for surface temperatures in the past, the ice sheet changed its size or shape greatly in the past, the model contains a mistake, the measurements contain errors, or something else—we really wouldn't know.

Fortunately, the isotopes predicted the borehole profile almost perfectly for both U.S. and European data. Surprisingly, the calibration was quite different from that obtained in 1964 by Dansgaard; the temperature in Greenland has changed twice as much as one would have calculated using the measured ice-isotopic shift and the modern relationship between temperature and isotopic ratio. Kurt showed that the coldest parts of the ice age were approximately 40°F colder than today on the surface in central Greenland, with an uncertainty of only about 2°F. That is a huge change, much larger than the 25°F difference in modern temperatures between frosty Chicago and toasty Miami in the United States. We believe that the calibration for the end of the ice age differs from Dansgaard's modern results because past changes occurred both in the temperature and in the season when most of the snow fell. During the cold times of the ice age, the winters were especially cold but especially dry, and without much wintertime snowfall, the ice isotopes didn't sample the full extent of the cooling.

When we discuss abrupt climate changes later, we'll come back to paleothermometry with some techniques based on gas isotopes. These techniques reveal temperatures only at certain times of abrupt change, but can yield quite accurate temperature reconstructions. The temperature estimates from these techniques agree beautifully with the estimates from ice isotopes and borehole temperatures. The conclusion is clear—we know how cold or warm the climate was in Greenland's past.

7

Whatever blows around in the air can land on an ice sheet and be buried in the snow. We can then analyze those materials and learn the history of things in the atmosphere.

Aerosols—small particles in the air—have many sources. Dust blows from continents, and especially from the great deserts of the world. Dust from the Sahara falls on the Americas, and Chinese dust helps build soils in Hawaii. The details of the chemistry and minerals of dust can be used to learn where it came from. Such studies show that much of Antarctica's dust blows in from Patagonia in southern South America. Amazingly, such studies also show that much of Greenland's dust is from Asia and has blown halfway around the world to reach the ice sheet.

Processes on land make other materials. The blue haze over the Great Smoky Mountains of North Carolina and Tennessee includes a variety of chemicals that are produced by trees and soils. Traces of these chemicals blow over Greenland and fall out in the snow. Forest fires put concentrated pulses of soot, ammonium, organic acids, and other chemicals into the air, and snow that falls through this smoke can collect enough of the smoke to form a recognizable layer in ice. Some pollen serves its reproductive function, some is filtered out of the air in the noses of allergy sufferers, but much pollen blows away to fall in lakes, on oceans, or on ice.

The ocean puts huge volumes of salt and other things into the air. If you hold your hand over a newly poured glass of cola, you will feel the little drops thrown into the air by breaking bubbles. If you allow the cola on your hand to evaporate, you will find brown, sticky spots, showing that the bubbles were throwing sugar and artificial coloring out of the glass. When waves break in the ocean, they turn white as they trap air and form bubbles; when those bubbles rise to the surface and break, they throw droplets into the air in the same way that soda bubbles do. Those droplets carry water, salt, and occasionally tiny shells and organic matter from living things. Hence, the air is constantly receiving ocean samples, and some of those end up in snow.

Large volcanic eruptions can propel cubic miles of material into the stratosphere, and this material may remain up for a few years before gradually falling out. Ash particles and sulfuric acid in ice cores are recognizable markers of the fallout of such globe-girdling eruptions, or of smaller, more local eruptions.

The atmosphere is chemically more interesting than anyone imagined even a few years ago. All sorts of reactive chemicals are supplied to the atmosphere from land and sea, and the sun's energy takes some of these chemicals apart, puts new chemicals together, and otherwise stirs the pot. One example is hydrogen peroxide, a chemically active compound that we humans use to sterilize cuts and scrapes, and that sunlight makes in small quantities in the atmosphere. Because hydrogen peroxide doesn't remain in the atmosphere very long, and because the sun doesn't shine to produce peroxide during the winter in polar regions, hydrogen peroxide has a beautiful annual signal and has been used in some ice-core dating, as noted above. The peak concentrations of peroxide in summer snow depend on many things, including the concentrations of other chemicals in the atmosphere with which the peroxide can react, so hydrogen peroxide concentrations can tell us something about the history of atmospheric chemistry.

Cosmic rays from space are constantly bombarding the

atmosphere and the rocks, snow, and water near Earth's surface. The huge energies of cosmic rays allow them to occasionally split an atom in the air or on the surface. Be-
ryllium-10 and other isotopes that are made in the
atmosphere by such splitting then fall on the surface, including on the ice sheets. If accurate ages are available, and certain other corrections are made for changes in dilution by
snowfall, as described soon, the concentrations of these cosmogenic types can be used to learn changes in their production rates. Production-rate changes can be caused by changes
in the cosmic ray flux, or in Earth's magnetic field, or in the
sun's activity (because the magnetic field and the solar wind
block many cosmic rays from reaching Earth). By comparing
ice-core data with results from other types of records, we can
learn something about Earth-space interactions.

Occasionally, one finds a meteorite, a micrometeorite, or
other space debris in the ice. Meltwater ponds on the edge of
the Greenland ice sheet often contain deposits of micrometeorites. In Antarctica, intense winds evaporate ice flowing
to the surface in a few special places near mountains, and
meteorites are concentrated on the surface as the ice is removed around them. The South Pole Station in Antarctica
uses waste heat from its electrical generators to melt ice for its
water supply. A water-filled cavity is maintained beneath the
station, slowly getting bigger and deeper as heat is added,
and enough water is drawn off to supply the station. The micrometeorites in the ice are released by melting, and are left
resting on the ice at the bottom of this cavity. A team of researchers has now built a robotic "vacuum cleaner" to collect
these small meteorites. If the infall of micrometeorites has
changed over time, then the slow downward melting in the
water well will cause the vacuum cleaner collections to
change over time.

In short, a clever person can find much in ice that is interesting to study. Many clever people are studying things in ice,
and learning many things. There are, however, a couple of
complications. One is that ice is still a rather clean material—
some of the interesting contaminants are measured in parts

per million in the ice, others in parts per billion or even less. One nose drip has more of some contaminants than do armloads of ice cores. Extreme care is required in collecting, handling, cleaning, and analyzing ice cores so that we learn about past climate conditions and not about the dirtiness of modern gloves. Ice chemists, especially those working on mercury, lead, and other naturally rare materials that humans have made common, often seem a little strange and paranoid to "normal" people. However, "just because you're paranoid doesn't mean they're not out to get you"—the dirty environment made by humans really is out to get ice chemists, so they need a little paranoia.

Another complication is that an ice sheet is not the perfect way to sample the atmosphere for contaminants. The processes by which materials are transferred from the air to the ice sheet can be quite complex. The concentration of a contaminant in snow is related to its concentration in air from which the snow fell, but snow concentrations may be related to other things as well.

For example, wind can move through the upper few feet of snow, although not much deeper than that. Wind blowing into the side of a snowdrift may carry a lot of little particles, but the snow acts as an air filter, and wind coming out the other side is much cleaner. If the climate became windier, the snowdrifts might become dirtier because more air would flow through.

Most of the chemicals that end up in snow or ice are collected in the atmosphere by snowflakes, either because the snowflake grew around a particle or because the falling snowflake collided with and stuck to the chemicals. The transfer of chemicals to the ice sheet in snowflakes is called "wet deposition" (even though, in our case, the "wet" is frozen). But some of the chemicals in snow fall right out on the snow surface or are blown into the snow by wind, in what is called "dry deposition."

Having both wet and dry deposition creates a bit of a problem—if not much snow falls, then the chemicals deposited dry will not be diluted by much frozen water, and the dirt concentration in the ice sheet will be much higher than that in falling snowflakes; if more snow falls, then the dry deposition will not be very important, and the dirt concentration in the ice will be almost the same as that in falling snowflakes. Fortunately, we have a fairly good way to learn how much dry deposition occurred and how dirty the air was.

Separating dry and wet deposition requires that an ice core be well dated and that the flow-thinning of the layers be known so that the rate at which the snow accumulated can be calculated. More chemicals reach the surface during snowy years because the extra snowfall brings down more chemicals from the atmosphere. In years with less snowfall, the chemical supply is smaller. But, as the snowfall drops closer to zero, the chemical supply does not drop toward zero—it drops toward the dry-deposition rate. If we plot the rate at which chemicals reach the surface against the snow supply, a line through the points on the graph can be extended to the zero-snow level, where the line will give the dry-deposition rate.

This allows us to separate dry deposition from wet deposition, and to learn how the wet deposition is related to the snowfall rate. Dirtier air makes dirtier snowflakes, and dirtier or windier air makes more dry deposition. We have a pretty good idea how to tease all of these apart. With a little effort, the ice core tells us how much sea salt and continental dust were blowing around, how many fires were occurring upwind, how well we were shielded from cosmic rays, how many meteorites were being dumped on Earth, and much more.

8

Most glacier ice is bubbly. As snow turns to ice, much of the air between the ice grains is squeezed out, but some is trapped as bubbles. Ice is a great material for bottling old air—gas molecules don't interact with the ice much, and gas molecules have a great deal of difficulty moving through the ice. The bubbles thus contain samples of old air, stacked up by age.

Old air is a difficult thing to find except in ice cores. Some researchers have looked in antique fishing floats and the artwork of glassblowers, hoping to learn how the atmosphere has changed. Unfortunately, this tells more about conditions in factories and in the lungs of glassblowers than in the natural atmosphere.

The atmosphere does change over time; those changes reflect processes that are occurring on Earth. The atmospheric changes also help to control the climate.

The winds mix the atmosphere quite well in just a few years. If you released a large amount of some gas in your backyard today, and that gas survived chemical processes in the atmosphere, in a few years everyone on Earth could measure what you had released. Dust and sea salt typically stay in the atmosphere for only days or weeks, shorter than the time for global mixing, so the dust falling on Antarctica is different from the dust falling on Greenland. But most gases stay in the atmosphere long enough to be mixed globally.

Because of the global mixing of most gases, one check on the reliability of ice-core gas measurements is to compare results from different cores from Greenland, Antarctica, and high mountain glaciers. With certain well-understood exceptions, such comparisons show that the ice-core data are quite reliable. These comparisons have worked so well, and the ice-core data have proved so reliable, that we now can use ice-core gases to correlate cores. The well-dated Greenland ice cores have the same gas history as Antarctic or high mountain cores that are harder to date, so one can assign ages determined in Greenland to Antarctic or mountain glacier samples based on the gas composition. Studies of gases that have been trapped in ice cores during the time when humans have been measuring the atmospheric composition directly show wonderfully good agreement between ice-core and atmospheric measurements, providing further confidence in the ice-core data.

Methane and carbon dioxide are two important greenhouse gases. Before the industrial revolution, methane was produced primarily in the swamps of the world, and its concentration in ice-core bubbles tells how widespread these wetlands were. Before the industrial revolution, carbon dioxide changes over tens of thousands of years were controlled primarily by oceanic processes, so carbon dioxide concentration in the atmosphere was a tracer of ocean chemistry. Both carbon dioxide and methane were low during the cold ice ages, and increased as the world warmed, with important implications for the way the climate works.

Freons that we humans have manufactured and released to the atmosphere have participated in some complex chemistry that leads to breakdown of ozone. Because ozone is important in shielding life on Earth from ultraviolet rays that can cause cancer and other harmful effects, humanity has decided to replace the more harmful freons with safer chemicals. Analyses of old gases in ice cores show clearly that freons were not around before humans started making them—thus, we really are responsible.

We will leave ice-core gases here with this very brief introduction. But we will come back to their important records several times as we try to understand the past—and future—of the Earth system.

CRAZY CLIMATES

What happened

to Earth's

climate in the

past—and

some ideas on

why the older

changes

happened

9

By now, I hope that you are convinced that a dedicated team of drillers, pilots, cooks, scientists, and others can pull a two-mile-long piece of ice out of Greenland, cut up the ice, analyze it, and tell you how and when the climate changed in Greenland and in many other places. Our friends can analyze trees and mud from other regions, and tell you much about the past climates where the trees grew and the mud settled. The stories from these studies, and what they might mean, are the reason the government paid for us to go to Greenland, and form the rest of this book. I'll give you the punch lines first, and then discuss them. There are many punch lines, and all have something to tell us. The two biggest are:

1. Climate in the past has been wildly variable, with larger, faster changes than anything industrial or agricultural humans have ever faced.
2. Climate can be rather stable if nothing is causing it to change, but when the climate is "pushed" or forced to change, it often jumps suddenly to very different conditions, rather than changing gradually. You might think of the climate as a drunk: When left alone, it sits; when forced to move, it staggers.

Some other interesting results include:

3. The "pushes" that have caused climate changes in the past probably have included drifting continents, wiggles in Earth's orbit, surges of great ice sheets, sudden reversals in ocean circulation, and others.

4. Small "pushes" have caused large changes because many processes in the Earth system amplify the pushes. Greenhouse gases have probably been the most important amplifiers.

5. Humans can foul our own nest—and we can clean it up.

Some of these results and discoveries are of academic interest, some may help us predict the future, and some may help us decide how to behave in the future.

To understand these results, we will start way back in the past. Changes in Earth's climate over millions of years have not been too much larger than changes over mere years. However, the "pushes" or "forcings" that caused the changes over millions of years have been much larger than those acting over shorter times. The difference must be in feedbacks—all of those processes that magnify or shrink the climate response to a forcing. Over long times, Earth's feedbacks act to oppose the forcing, so large causes produce small effects. Over shorter times, Earth's feedbacks amplify the forcing, so small causes have large effects.

Through the four billion years of Earth history that we can read in sedimentary rocks, changes in the sun's brightness have been nearly offset by changes in the greenhouse gases in Earth's atmosphere. Climate has changed over hundreds of millions of years as continents drifted over the surface of the globe, changing ocean currents and winds, and changing the rates at which greenhouse gases were added to the atmosphere by volcanoes and removed from the atmosphere by reactions with volcanic rocks.

Over hundreds of thousands of years, colder and warmer times have oscillated as tiny wiggles in Earth's orbit moved sunlight from place to place and season to season on Earth.

The great ice sheets that grew to cover much of Europe and North America only 20,000 years ago were responding to these orbital effects. Oddly, the whole globe was cooled during this ice age, even though half of the globe received more sunshine than today, and even though the total sunshine received on Earth was essentially the same as today. Again, greenhouse gases are implicated.

The coming and going of ice ages and other slow changes tell us much about the climate, but these changes are not fast enough to really affect us, and they will not do much to counteract faster changes caused by nature or humans. After we spend the next three chapters looking at the "how" and "why" of slow climate changes, we will finish this section with the remarkable evidence for jumping climates, which have been especially prominent during coolings and warmings of the orbitally driven ice ages. These abrupt changes are so startling that we will save their causes for Part IV.

The Faint Young Sun

Ancient records of climates from the deep past, much older than ice cores, tell of large changes that fortunately haven't been too large. Surprisingly, climate changes over millions of years weren't too much bigger than changes over years to decades.

At least some of the sedimentary rocks from all Earth's geological ages have features showing that the rocks started as sediments deposited in liquid water. Throughout four billion years of history, the surface temperature has not strayed far enough from comfortable conditions to turn us into a Martian deep freeze or a Venusian broiler. This temperature stability is quite surprising. Based on our understanding of solar physics, the sun has slowly been getting hotter, and it supplied only three-quarters as much heat four billion years ago as it does today. Why didn't we freeze completely back then, or why aren't we broiling now?

The generally accepted solution to this "faint young sun paradox" is that we were saved by chemistry. Carbon dioxide is a greenhouse gas, which catches and sends back to Earth some of the heat that we otherwise would lose to space.

Rainfall picks up some carbon dioxide from the atmosphere, making a weak acid. That acid attacks rocks and breaks them down, in a process called *weathering* because it is driven by the weather. The chemicals weathered free from the rocks are washed to the ocean. There, corals and other creatures use some of the chemicals to grow shells. We can write an equation for this long, drawn-out chemical reaction: rock ($CaSiO_3$) plus carbon dioxide (CO_2) combine to form carbonate shells (corals, clams and others; $CaCO_3$) plus silicate shells (sponges, diatoms, and others; SiO_2). Eventually, processes associated with the drifting of continents drag the shells deep into hot regions of Earth. There, heat reverses the weathering reaction, making melted volcanic rock and carbon dioxide gas that are both erupted from volcanoes to allow rock weathering to begin again. (I have used a convenient and oversimplified chemical formula to represent "rock" here. Rocks also contain other chemicals, including iron that rusts and feldspars that form clays, contributing to soil formation. Soil is also washed to the sea, dragged down, melted, and erupted. And metamorphic reactions can remake rock and release carbon dioxide without actually melting.)

The rate at which volcanoes bring carbon dioxide and volcanic rocks from deep inside Earth back to the surface depends on deep-Earth conditions that change only over millions of years. But the rate at which carbon dioxide is removed from the atmosphere depends on temperature, because higher temperatures cause chemical reactions to run faster. If the climate cools, the chemical reactions between carbon dioxide and rocks slow down, the volcanic supply of carbon dioxide to the atmosphere exceeds weathering removal, carbon dioxide increases in the atmosphere, and the world warms up. If the climate warms too much, carbon dioxide reacts with rocks more rapidly than volcanoes replace it, so atmospheric carbon dioxide levels drop and the climate cools off. Nature has given Earth a thermostat that keeps the surface habitable for creatures such as us, who like liquid water.

This thermostat is an example of a stabilizing process that tends to "fight" against some change to make the change smaller, and is called a negative feedback. We are most familiar with feedback from cheap public-address systems at county fairs and similar gatherings. A technician turns on the system and taps a fingernail on the microphone. An amplifier in the system makes that fingernail tap a lot louder, and sends the noise out from the speaker, which the technician has mistakenly placed right next to the microphone. This loud noise from the speaker is picked up by the microphone, and sent back out of the speaker even louder. The result, a nasty whine, is a positive feedback—the initial noise of the technician tapping on the microphone triggers other processes that magnify the initial noise.

Your body makes good use of negative feedback processes. For example, if you exercise, you begin to warm up. But you then start to sweat, and the evaporation of the sweat cools you off. Should you become too cool, you will stop sweating, allowing you to retain more of your body heat, and you might even begin to shiver to make more heat.

You have positive feedbacks as well. Fever helps your body fight invading germs. When you have a fever, your body convinces you that you are still cold, and you pile on more blankets to warm yourself even more. In some cases, such positive feedbacks may get "carried away," and a body may warm itself until it is damaged.

The Earth system is full of feedbacks, too. The longest-term feedbacks clearly are negative, and have served to stabilize Earth's climate for four billion years in that narrow temperature range where liquid water occurs. We will see that many of the shorter-term feedbacks are positive, causing Earth's climate to vary almost as much over years, decades, or millennia as it has over billions of years.

The changes in the sun's brightness, and the response from rock weathering, are slow processes. If you worry about times as short as a few million years, the solar physicists do not believe that the sun's brightness has changed very much. Changes in interactions of rock and carbon dioxide can affect Earth's temperature significantly in half a million years or less, but big changes usually require millions of years or more. These sorts of processes won't help humans much in dealing with changes that happen on "our" time scales of civilizations, lifetimes, or congressional terms.

The geological records that show climate stability over billions of years also show that the amount of ice on Earth has changed greatly over hundreds of millions of years. Today, permanent ice covers 10 percent of the land surface, but 100 million years ago, there apparently was no land ice on a much warmer Earth. During that Cretaceous warmth, dinosaurs stomped around steamy swamps or swam in warm oceans without worrying about running into glaciers and icebergs. Further back, many such warm and cold periods have occurred. Some geological evidence suggests that a few of these old, cold periods may have come perilously close to turning the planet into a giant "snowball Earth" before the slow supply of carbon dioxide from volcanoes saved us. The time it took Earth to go from warm to cold or back again has varied quite a bit, but typically has been closer to 100 million years than to ten million years or to a billion years.

It may come as no surprise that 100 million years is about how long it takes for continental drift to really change the appearance of the Earth—not much change happens in ten million years, and continents can be rearranged several times in a billion years. If all the continents sit on the equator, then there is no cold land on which to grow ice sheets; if a continent sits isolated on a pole, such as Antarctica on today's South Pole, then glaciation is easy. Drifting continents also affect whether the carbon dioxide-spewing volcanoes are common or rare, what rocks are exposed to react with the carbon dioxide, and how the currents of the oceans and atmosphere are arranged. So the typical hundred-million-year spacing between icy and hot periods on Earth is the typical time for continental drift to "shuffle the deck" and create a new configuration of continents, mountain ranges, ocean currents, and so forth. We are living in one of the colder times, millions of years with much ice and not much carbon dioxide. We are near the bottom of the 100-million-year slide from the saurian sauna of the Cretaceous.

As an aside, the meteorite that is believed to have killed the dinosaurs probably did so in part by changing the climate. Strong evidence shows that a meteorite impact about sixty-five million years ago blasted rocks out of a gigantic crater on the tip of the Yucatan Peninsula. The bigger rocks were tossed out of Earth's atmosphere in huge quantities and spread around the world. These rocks then heated up like meteorite "shooting stars" as they fell back, seconds to hours after the impact, starting fires that raged across much of the land surface. Dust particles and vaporized chemicals from the impact would have fallen so slowly that they did not heat up, but stayed aloft for years blocking the sunlight. The cold of the resulting "impact winter," close on the heels of the devastating fires, would have been much too stressful for many living things.

But the climate seems to have jumped back to a normal state within decades or centuries. The meteorite is estimated to have been less than one-billionth of the mass of Earth, not nearly large enough to change Earth's orbit significantly or roll the planet over on its side or anything like that. The climate probably didn't quite get back to where it was before the impact because living things do affect climate, and the types and distributions of living things did change after the impact. However, the climate changes from the biological effects of the impact were probably rather small compared to the vast changes from the saurian sauna to the recent ice ages.

The study of great changes over deep time is one of the great joys of geology, and has much to teach us about the Earth system. But catastrophes such as the dinosaur-killing meteorite are, fortunately, rare—nothing nearly so large has happened in the intervening sixty-five million years. The other slower changes of deep time serve as the stage on which faster events occur. We move next to the faster changes of the ice ages, and then to the frenetic changes written in the Greenland ice cores.

10

Earth appears to have had a significant amount of ice for at least the last few million years, so we are away from the warm end of the full range of Earth climates. But the climate has been far from static over these few million cold years. Ice covers 10 percent of our land today, but covered 30 percent of our land only twenty thousand years ago. Walk across the U.S. Midwest or the plains of northern Europe and you'll find the deposits left by vast ice sheets. These deposits are layers of mud and rocks that the glaciers picked up as they moved, and then spread across the land surface the way a knife spreads peanut butter across bread. Stacks of these deposits show that the ice has come and gone many times.

However, if you think about spreading layer after layer of peanut butter on bread, you can imagine that sometimes two layers become mixed, or the knife will grab several layers at once and move them, eroding down to the bread. Also, if your three-year-old child were nibbling on the edge of the bread even as you tried to spread the peanut butter, you would lose some of the record of your spreading, just as the streams fed by melting ice remove much of the glacial record on land by carrying the loose rocks from the glaciers to the sea.

To learn how the ice has grown and melted on land, we follow those streams to the sea. There, we will find a "dipstick" to show how much water was left in the oceans and how much was locked up as ice on land at different times. Using some elegant mathematical techniques to read this dipstick shows that features of Earth's orbit have caused the ice to grow and shrink in response to changes in the summerwinter, north-south distribution of sunlight. Surprisingly, we will find that sunshine reaching the high northern latitudes of Canada, Europe, and Siberia is more important in controlling the world's climate than sunshine in other areas.

A Deep-Ocean Dipstick

Collect a sediment core from the ocean floor, and you will find that the mud contains the shells of tiny plants and animals that lived in the water or on the sea floor and piled up in layers over millions of years. Those shells are often made of calcium carbonate or silica, which contain oxygen that the creatures extracted from sea water.

If you recall from our discussion on measuring past temperatures, oxygen naturally has different "flavors," or isotopes with different weights. Lighter isotopes have an easier time evaporating from the ocean than do heavier ones, so water vapor and rainfall are isotopically lighter than the ocean. A glacier or ice sheet is just a great pile of water vapor turned into snow and stored on a continent somewhere. Today, all of those piles contain enough water to raise the level the world's oceans more than 200 feet vertically if melted; 20,000 years ago, the ice held an additional 400 feet of sea level.

When the ice sheets were bigger, they contained much more isotopically light water that the ocean had lost. Heavy isotopes were concentrated in the water remaining in the ocean. The plants and animals that obtained oxygen from the ocean to grow their shells had to use more heavy oxygen than usual in those shells. When the ice sheets melted, the isotopically light water flooded back into the oceans, where it was used to make isotopically lighter shells.

So, as younger shells have piled up on old ones, they have been writing a history of ice volume. Take a drill ship to sea, and pull up a core of the sediment. Pay some poor student or technician to sort through the mud and pull out the shells of your favorite "bug" type. Use some of the great range of dating techniques to assign ages to the shells. Run the shells through your local mass spectrometer to measure the ratio of heavy to light oxygen. The result is a record of the size of ice sheets on Earth.

There are, of course, a few complications. Some species "prefer" heavy or light oxygen, so you need to be sure that all of your shells are from the same species, and to use a species that is known in the modern world to record the water isotopes accurately. Temperature, as well as water composition, affects the isotopes taken into a shell, but the water-composition effect is usually much greater, and there are ways to correct for temperature. Suffice it to say that one can turn the history of isotopic compositions of "bug" shells into the history of ice volume on Earth with some confidence.

Questioning Periods

This history of ice ages from bug-shell isotopes appears quite regular over the last million years (see Figure 10.1). Ice grew for about 90,000 years, shrank for about 10,000 years, and repeated, with smaller wiggles spaced about 19,000 years, 23,000 years, and 41,000 years apart. These numbers are obtained from application of Fourier analysis, developed by the French mathematician of that name.

To see how this works, suppose that you are a weather wonk living in the midlatitudes, such as in my home in central Pennsylvania. Every hour, day and night, you write down the temperature from a thermometer sitting outside of your window. After a few years, you take your pages and pages of numbers and start to analyze them. If you drew the wiggly line of how temperature changed over time, you would see several things: Days typically are hotter than nights; summers typically are hotter than winters; the weather often warms for a few days, cools as a cold front storms through, and then warms again; and the data are "noisy," in that some tempera-

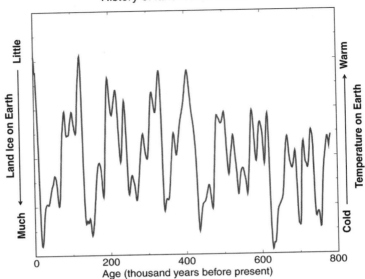

FIGURE 10.1

The history of ice volume on Earth (with more ice toward the bottom of the page), which is also approximately the history of temperature on Earth (with colder conditions toward the bottom of the page), from the work of John Imbrie and others as cited in the Sources and Related Information. Especially cold, large-ice times are spaced roughly 100,000 years apart, with slow, bumpy, 90,000-year coolings leading to them and faster, 10,000-year warmings ending them. The spacings of these warm and cold times have been controlled by the wiggles in Earth's orbit.

ture changes cannot be explained by any of the daily, weekly, or yearly changes. You could do a reasonably good job of inventing a realistic-looking temperature record if you simply took cold-winter/hot-summer plus cool-night/warm-day plus a week-long storm cycle, and then added or subtracted some random numbers to give an appropriate, messy appearance.

Many real systems look like this. The variability in something of interest, such as temperature, depends on some "clocks," such as the day-night and summer-winter changes. The variability also includes some "characteristic times," such as the week between cold fronts, which is not a perfect clock because cold fronts could be three days or two weeks apart, but which has a "favorite" or "characteristic" spacing. Fourier developed the mathematical techniques, now routine, for finding out what sorts of variability are contained within any data set. Applying these techniques to the few years of temperature records from outside your window would produce clear daily and yearly "peaks," as well as some variability centered on the week-or-so storm cycle. You would say that there are daily, weekly, and yearly periodicities, named for the time or period over which each repeats.

These same techniques have been applied to the history of ice volume from ocean shells over the last million years. Before the 1990s, most ocean-core studies used sediment from places where worms and other burrowing animals had crawled around and stirred up the mud while looking for something to eat or hiding from things that wanted to eat them. The typical effect of this biological perturbation, or "bioturbation," is to smear out the records of any changes lasting a few thousand years or less, but to leave longer records intact. You could create a similar effect if you graphed your hourly temperature data using a soft lead pencil, and then smeared your finger across the page—the day-night changes might disappear, although you could still see the slower changes. Study of the worm-stirred records found three main spacings or periodicities, with peaks in ice volume separated by about 100,000 years, 41,000 years, and a sort of doublet at 19,000 and 23,000 years.

Celestial Seasonings

Remarkably, these periodicities were predicted decades before they were observed. A Yugoslavian mathematician, Milutin Milankovitch, had learned from earlier workers that Earth's orbit has some odd features because of influences such as the gravitational tug of Jupiter. These orbital features have little effect on the amount of sunlight that Earth receives, but they do change where on Earth, and during what season, the sunlight is received (Figure 10.2).

Earth's rotation axis (the line connecting the north and south poles) is inclined to the orbital plane. If, each day, you could somehow place a stick from the sun to Earth, after a year those sticks would make a sort of orbital table or plane. Earth's rotation axis doesn't stick straight up from that plane, but is angled about 23° away from "straight up." This is why we have seasons—the northern hemisphere is tipped toward the sun on one side of the orbit, and away from the sun on the other side of the orbit, while the opposite is true of the southern hemisphere. If the rotation axis stood "straight up," the poles would be in perpetual twilight rather than switching

FIGURE 10.2

The features of Earth's orbit that redistribute sunshine on Earth and drive the ice-age cycles. The "out-of-roundness," or eccentricity, of the orbit increases and decreases with a 100,000-year period. The spin axis connecting Earth's North and South Poles is inclined to the orbital plane, and the "tilt" or inclination of the spin axis changes with a 41,000-year period. The "wobbling" of this spin axis causes the time when Earth is closest to the sun to shift, or precess, from northern-hemisphere summer/southern-hemisphere winter (as shown) to southern-summer/northern-winter and return, with typical spacings of 19,000 to 23,000 years. These orbital features don't have much effect on the total sunshine received by the planet, but they change where and during what season the sunshine is received.

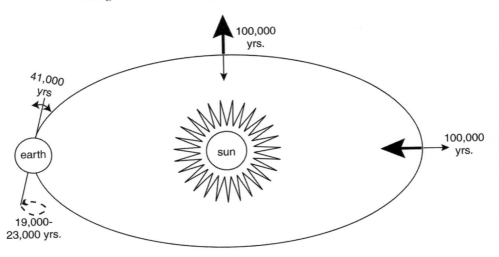

from 24-hour daylight to 24-hour darkness, and the seasons as we know them would not exist. The angle of the rotation axis, called its *inclination,* or *obliquity,* varies from about 22° to 24° and back over about 41,000 years. As the inclination changes, the difference between summer and winter temperatures changes. A decrease in inclination toward the "no-inclination, no-seasons" case causes the difference between summer and winter temperatures to become smaller, both in the northern hemisphere and in the southern hemisphere, giving warm rather than hot summers and cool rather than cold winters toward both poles. The difference between summer and winter temperatures increases at both poles as the inclination increases.

Planets run in oval (elliptical) orbits rather than circles. The *eccentricity,* or "out-of-roundness," of Earth's elliptical orbit about the sun varies, going from more nearly round to more flattened and back over about 100,000 years. In addition, the rotation axis wobbles slowly, changing the season when Earth is closest to the sun or farthest from the sun on the elliptical orbit. This "precession" of the seasons has a complicated variation that repeats typically every 19,000 or 23,000 years. Today, the northern hemisphere has summer and the southern hemisphere has winter when Earth is far from the sun, with northern winter and southern summer when Earth is close to the sun. About 10,000 years ago this was reversed, with northern summer and southern winter occurring when Earth was closest to the sun. Thus, the modern summer-winter difference is larger in the south and smaller in the north than it was 10,000 years ago. Note that the main effect of the 100,000–year change in the out-of-roundness of the orbit is to determine how important the precession is—if the orbit were perfectly round, then Earth's distance from the sun would not change, precession wouldn't matter, and only the inclination would affect seasons. The orbit actually is nearly round now, so the precessional effects are not large.

From all this, Milankovitch calculated the periodicities of sunlight shifting across Earth. Decades later, the isotopic composition of small fossils in ocean sediment showed just those periodicities. The agreement is too good to be a coincidence; the changing sunlight must have been causing the ice to grow and shrink.

The highlands around Hudson Bay and in some other parts of the far north have average temperatures well below freezing, but the little snow they receive melts in the summer. Shorter, cooler summers would allow more snow to survive, and longer, milder winters might actually promote snowfall, because temperatures would still be below freezing but the warmer air could deliver more moisture. Comparing the ocean-core record of global ice volume to Milankovitch's calculations of sunshine shows that ice has grown when Hudson Bay and northern Europe had short, cool summers and long, mild winters, with ice melting when Hudson Bay summers were long and warm.

A couple of curiosities are related to these patterns of ice growth and melting. One is that ice grew in Antarctica, New Zealand, and Chile when Hudson Bay and northern Europe had reduced summer sunlight, even if Antarctica, New Zealand, and Chile were getting more summer sunlight than usual. We will come back to this bizarre behavior soon.

The other curiosity is that small changes in the distribution of sunshine over 100,000-year periods have caused big changes in ice sheets, while larger changes in sunshine over shorter periods have caused smaller changes in ice sheets. The strange explanation of this behavior is that a big ice sheet may melt more rapidly than a small one, and it takes most of 100,000 years to grow a big ice sheet.

We must now solve two riddles—why tiny changes in sunlight over 100,000 years have caused such huge changes in the world's climate, and why sunlight on Canada, Europe, and Siberia is more important than sunlight on New Zealand Antarctica, or many other places. Neither riddle has been fully solved. It seems as though the ice sheets themselves are responsible for the size of the 100,000-year changes, primarily because the ice sheets take tens of thousands of years to grow but only thousands of years to die. And, somehow, it appears that the levels of carbon dioxide and other greenhouse gases drop as northern sunshine dims and ice sheets grow, and rise as the northern sunshine increases and ice melts. These changes in greenhouse gases, in turn, cause temperature changes worldwide.

The Bigger They Are, the Faster They Fall

Why can a big ice sheet disappear more rapidly than a small one? At least three reasons probably are important: A bigger ice sheet has a warmer bed, has a warmer surface, and can flow into warmer water around it. We will work through these in turn.

Remember that a glacier or ice sheet is just a way to take snow from where too much falls to somewhere the excess can melt. All glaciers move ice by slow deformation within themselves. Some glaciers also move ice by sliding over their beds. A thick ice sheet acts like a blanket, trapping Earth's heat at the bottom of the ice and keeping the cold air away. If this blanket is thick enough, the ice at the bottom begins to melt. Then, the glacier or ice sheet can spread and thin more rapidly. Sometimes, this melting of the glacier bed can allow the ice to go more than one hundred times faster than if the ice were still frozen to the rocks beneath it.

A large glacier or ice sheet is extremely heavy, and pushes down the land beneath it and nearby. If you have ever sat on a water bed, you know that the water bed cover sinks beneath you as water is pushed out to other parts of the mattress. If the water bed cover is stiff enough, you will see that the surface is pushed down in a little dimple around your behind. When you stand up, the water flows back in and the surface of the bed rises where you had been. If the water bed were filled with maple syrup or gelatin, the sinking and rising of the surface would take a while, and you could watch the rising of the dent left by your behind.

The cold, stiff rocks near Earth's surface form a "water bed cover" sixty miles or so thick, resting on soft, hot rocks beneath. If you could suddenly drop an ice sheet on Earth, the rocks beneath the ice and nearby would sink. However, the sinking would take thousands of years because the flow of the hot, deep rocks is still rather slow. If you then melted the ice sheet suddenly, and at least most of the meltwater flowed away, the land would rise over many thousands of years as the soft, deep rock slowly flowed back. Today, careful measurements show that regions of Scandinavia and Canada that were pushed down by the ice-age ice sheets are still rising a fraction of an inch per year, to get back to where they started.

Snow-capped mountains show us that higher places are colder. Imagine a high plateau somewhere northeast of Hudson Bay, with a small ice cap sitting on top. Several such ice caps occur today in the Canadian Arctic islands. Snow that falls on top of these ice caps doesn't melt there, but feeds mountain glaciers that flow down off the plateau sides to lower, warmer elevations where they can melt. Such a small ice cap could sit happily on its plateau for a long time.

Now, suppose that the climate oscillates, getting periodically colder and warmer. During the cooling, the glaciers flowing off the side of the plateau do not melt. Instead, snow builds up on the plateau, the glaciers flowing off the plateau, and surrounding areas. If the cooling lasts long enough, this snow can accumulate to form an ice sheet two miles thick. The weight of this ice will depress the land beneath it into a great bowl, with the center pushed down more than half a mile.

The ice sheet thus grows on high land, which then sinks, lowering the surface of the ice to warmer levels of the atmosphere, where melting is more likely. At the same time, warming may occur as Earth's orbit rearranges the sunshine. Such warming can cause the edge of the ice to melt back rapidly, before the land can rise much, so the melting edge of the ice sheet will move back into a bowl of depressed land.

If the meltwaters from the ice somehow find a way to escape from the bowl, then the melting edge of the ice will move to lower and lower elevations into the bowl, where the temperature is warmer and warmer and so causes faster and faster melting. Steeper sides cause the glacier to spread and thin more rapidly, pulling its surface lower in the atmosphere, where melting is more rapid.

More commonly, the meltwaters will pond along the edge of the ice, or the ice will push the land below sea level and ocean water will flow in. Currents in oceans or lakes take waters heated by the sun and swirl the warm water past the edge of the ice. Just as ice cubes melt more rapidly in a glass of water than piled in an air-filled glass, marginal lakes or seas speed the melting of a glacier. Also, icebergs can break off in marginal lakes or seas and float away to melt elsewhere.

Thus, in a modern climate, a small ice cap might sit contentedly on a small plateau on a Canadian Arctic island almost forever. But if climate cooled for long enough to grow a really big ice sheet that pushed down the land beneath it, warming back to modern conditions could melt the ice sheet entirely before the land rose again.

One can build computer models of ice sheets, including all these processes, and ask what the known changes in sunshine would do to the ice. Many models agree that the ice grows for about 90,000 years, with wiggles in size spaced 19,000, 23,000, and 41,000 years apart. Then, the ice shrinks rapidly over about ten thousand years. Big peaks in ice volume thus are spaced about 100,000 years apart. We naturally should be near the start of the next long, slow, bumpy slide into an ice age.

So, we believe we understand why ice has grown and melted each 100,000 years recently. But why has ice grown in the south when the north had reduced summer sunshine, even if the south was having extra summer sunshine? Long ice-core records, especially the record from Vostok Station in central East Antarctica, clearly show temperature changes caused by orbital wiggles. Fortunately, the Vostok ice core also indicates one of the big reasons why: greenhouse gases. For reasons that we still do not fully understand, the cooling of ice-age cycles has reduced the greenhouse gases in the air, causing more cooling. Probably, water vapor fell when ice-age cooling reduced evaporation, and carbon dioxide fell when stronger ice-age winds blew more dust into the oceans to fertilize algae.

Gassing the Ice

Remember that the transformation of snow to ice traps bubbles of old air, which is held in the ice almost unchanged. This allows us to learn the history of greenhouse gas concentrations in the atmosphere.

The most important greenhouse gas is water vapor. The available evidence indicates that cooling reduced the water vapor in the air, which caused more cooling. Admittedly, sedimentary records of water vapor are not very good—sediment cores, trees, and ice all have lots of water all the time, so these sediments do not record subtle changes water vapor, but we can look to other information.

In general, warmer air holds more water vapor, so we would expect warmer times to have more water vapor. Available evidence supports this supposition. Almost all ice cores from polar regions and from high mountains show that snow accumulated more rapidly during warm times than during ice ages, and global wetlands seem to have been more widespread during warmer times. Certainly, temperature alone does not control precipitation, or the broiling Sahara would be the wettest place on Earth. Instead, both temperature and "storminess" contribute to precipitation, so a change in snow accumulation on an ice sheet shows a change in either temperature or storminess. But regions that are known to have been drier than today during ice ages are larger than regions known to have been wetter than today, and the wetter regions generally can be explained by changes in where storms typically went. Thus, it appears that colder times had less of the most important greenhouse gas, water vapor.

We can do much better in learning how other greenhouse gases changed. The next most important greenhouse gas is carbon dioxide, followed by methane, nitrous oxide, and others. These can be studied easily in the air bubbles in ice from Antarctica, and most are recorded well in ice from Greenland. (Some Greenland ice gives slightly anomalous carbon dioxide numbers because reactions between volcanic or other acids and carbonate dust have made a little carbon dioxide in the ice, but chemical contamination is much less of a problem in the cleaner ice of Antarctica. The chemical reactions that produce extra carbon dioxide do not produce methane, nitrous oxide, or some other important gases, so records of these gases from Greenland are fine.)

Studies on Antarctic cores show that greenhouse gases in the atmosphere fell during times when Antarctica was cooling (see Figure 11.1). These are the times when sunshine was decreasing in northern summers, ice sheets and glaciers were growing in much of the world, and sea level was falling. When summer sunshine rose in the north, melted the ice, and

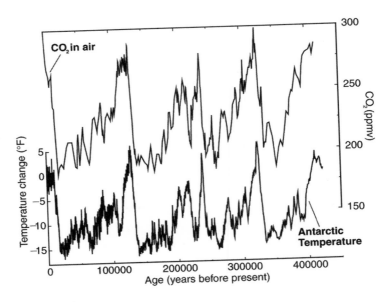

FIGURE 11.1

The ice-isotopic history of temperature in central East Antarctica at Vostok, and the history of CO_2 from air bubbles in the Vostok core, from the paper by J.R. Petit and others cited in the Sources and Related Information. This figure covers four cold ice ages. CO_2 levels have been high during warm times, and low during cold times, throughout this record. The temperature curve has not been successfully explained without including the effects of the CO_2 curve, providing strong support for the idea that CO_2 affects temperatures. The CO_2 changes probably were driven by orbital wiggles and their effects on ice sheets, wind, and other things, with the temperature then responding to the CO_2; however, the exact timing of the changes in CO_2 relative to changes in temperature has not been learned yet, as discussed in the next chapter.

refilled the oceans, greenhouse gases rose and Antarctica and much of the world warmed. All the greenhouse gases, including water vapor, seem to have changed more or less together, although with interesting small differences in timing. The ice-age world probably averaged about 10°F colder than recently, with some regions having cooled 40°F; much of this global temperature change is explainable by the changes in these greenhouse gases and associated feedbacks.

It is at once embarrassing and exciting to admit that while we do have a pretty good understanding of why the water vapor and methane changed (the discussion is coming), we do not really know why the carbon dioxide and the nitrous oxide changed. We have some good ideas, but we are not sure. Here, we will concentrate on carbon dioxide, because it is a much more important greenhouse gas than nitrous oxide. To understand carbon dioxide changes, we need to take a brief look at carbon on Earth.

Earth's atmosphere now contains about 800 gigatons—i.e., 800 billion tons—of carbon, almost all of which occurs as carbon dioxide. With 6 gigapeople—6 billion of us—on the planet, that amounts to about 133 tons of carbon per person. We are responsible for putting almost one-third of this carbon into the air, and we are likely to put even more carbon dioxide into the air in the future (see Figure 11.2). The world's forests hold almost as much carbon as the atmosphere does, and the organic matter in soils includes almost twice as much carbon as is in the atmosphere.

The ocean contains a huge amount of carbon dioxide that has dissolved and reacted with the water—about fifty times as much as is in the atmosphere. Even more carbon is locked up in carbonate rocks, but significant natural changes in this rock carbon occur only on the 100-million-year time scale of continental drift. Oil, coal, and natural gas that we humans can readily access contain about seven times more carbon than the atmosphere does.

Because the stores of carbon that can experience significant natural changes over millennia are dominated by the oceans, the carbon dioxide changes over ice-age cycles almost certainly depended largely on the oceans. True, the advance and retreat of glaciers must have changed carbon storage on land—if a glacier bulldozes a forest, trees are no longer growing there, for example—but these changes were small.

Several processes may have contributed to the oceans' absorbing carbon dioxide from the air during ice ages. One change is easy. Colder water holds carbon dioxide better.

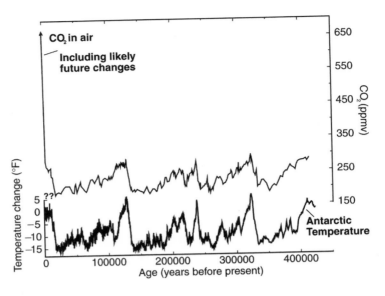

FIGURE 11.2

The history of temperature and CO_2 from Vostok, as in the previous figure, but with the scale changed to show what humans are likely to do to CO_2 within the next centuries. The question marks for future temperature pose some interesting questions for us.

Heating a can of soda will drive off the "fizz," which is just carbon dioxide. Heating the oceans does the same thing. Cooling the oceans during an ice age will cause them to absorb some more carbon dioxide, leading to more cooling.

Other changes are a bit more complex. When algae grow well in the surface ocean, they use energy from the sun to combine carbon dioxide with water, making more algae. Animals fuel themselves by eating and slowly burning algae or algae-eating animals, turning the algae back to energy, water, and carbon dioxide. However, animals are not perfect combustion chambers, and cannot use everything they eat. Instead, animals package some of their food into fecal pellets ("worm droppings") that sink quickly into the deep ocean and perhaps all the way to the sea floor, carrying some carbon and other chemicals with them. The carbon dioxide-algae-animal-droppings chain causes some carbon dioxide to drop into the deep ocean, lowering the carbon dioxide content of the surface ocean. More carbon dioxide then diffuses from the atmosphere to the surface ocean to replace the sinking carbon. Thus, biological activity in the surface oceans acts as a pump to transfer carbon dioxide from the air to the deep sea. And several processes may have caused this "biological pump" to work better during ice ages.

For example, during an ice age, the spread of ice and snow in polar and midlatitude regions causes more of the sunshine reaching the high latitudes to be reflected back to space rather than being absorbed to warm the planet. The ice-age cooling thus is larger toward the poles than in the tropics. Increasing the temperature difference between equator and pole drives the winds more strongly, and stronger winds pick up more dust from the continents and carry that dust farther over the oceans before dropping it. Windblown dust is a source of nutrients that fertilize the surface ocean. Ice-age cooling thus would have accelerated the biological pump that transfers carbon dioxide from the air to the deep ocean, lowering atmospheric carbon dioxide and causing even more cooling.

The biological pump also may have been affected by changes in sea level, caused mostly by growth and decay of the northern ice sheets. In the modern warm time, ice sheets are small, sea level is high, and our coasts are indented by numerous bays, such as Chesapeake Bay, San Francisco Bay, the Bay of the Seine near Le Havre, France, and the river Thames below London. These bays are drowned river valleys. They were carved during ice ages when sea level was as much as 400 feet lower than today and the sea had retreated from our modern coasts, often by tens or hundreds of miles. The nutrients carried by modern rivers tend to be trapped in sediment piles near the heads of these bays. During the ice age, more of the river-carried nutrients would have been delivered to the open ocean, perhaps fueling the biological pump and pulling carbon dioxide out of the atmosphere.

There are several other possible explanations for the ice-age changes in carbon dioxide. The details of the timing of temperature and greenhouse gas changes seem to require that more processes were active than we have discussed here. However it happened, we know that carbon dioxide levels fell as ice grew on Canada and Europe, and that this helped cool the entire world.

Certainly, carbon dioxide was not the whole story. After all, ice-age cycles were caused not by changes in carbon dioxide, but by orbital wiggles. Many factors other than carbon dioxide affected climate. Some temperature changes have occurred without carbon dioxide changes, and some carbon dioxide changes have occurred without temperature changes because other factors were more important. But carbon dioxide has been important, and there is no good way to explain the ice-age cycles without appealing to the importance of the carbon dioxide.

These ice-age cycles, with their huge changes over tens of thousands of years, have certainly succeeded in capturing the human imagination. It is only a geological eyeblink since New York and Stockholm were buried under huge piles of ice while mammoths walked the land around the ice. But on the time scale of human economies, the ice ages were "forever" ago. We will next look at changes that were fast even when compared to changes in our societies.

12

We have now seen that Earth's climate didn't change too much over billions of years, and that ice-free and icy times have alternated over hundreds of millions of years, leaving us in an icy time during which glaciers have grown and shrunk over the last hundreds of thousands of years. The most important processes have been different over these different time scales—changes in the sun have been offset by changes in greenhouse gas consumption by rock weathering over billions of years, continental drift has altered patterns of atmospheric and oceanic circulation and of greenhouse gas production and removal over hundreds of millions of years, and features of Earth's orbit have affected the distribution of sunlight, ice, and greenhouse gases over hundreds of thousands of years.

All these changes share one important characteristic—they aren't occurring fast enough to matter directly to your grandchildren. I think that they are interesting. We know that they tell us a lot about the Earth system, but they are "given" for the next centuries or millennia. Just as building a stadium is critical to the long-term success of a sports franchise but doesn't indicate the outcome of a game, building our climate determines what is and is not possible, but it doesn't tell what actually will happen.

In this chapter, we will begin to consider things that have

happened fast enough to matter to you, as well as to your grandchildren. We then will use the three chapters of the next section to try to explain these abrupt changes, pointing toward predictions for the future in the last section of the book.

Something Happened Then

Most of the records of the slow, orbitally caused climate changes came from ocean sediments that had been stirred by burrowing creatures, smoothing out the history of faster changes. The long ice-core records from Antarctica have come mostly from places where the average snowfall that accumulates each year is thinner than the height of a snowdrift, so that annual layers are not preserved.

Certain marine sediments preserve fine details, and researchers studying them now are finding a wealth of information on rapid climate changes. Records from land—pollen in peat bogs, tree-ring widths, and such—also have information on rapid climate changes. The search is on for long, annual records from Antarctic ice. But the clearest story of these changes, of how big, widespread, and rapid they were, has come from the high-resolution ice cores of central Greenland.

At the beginning of this book, we met the Younger Dryas, the last cold gasp of the ice age between about 12,800 and 11,500 years ago. *Dryas* is a pretty little mountain flower of the rose family, white with a yellow center, often called avens, that today lives in high-altitude or high-latitude cold places. If you collected a sediment core from beneath a bog in a European forest today, you would not find evidence of *Dryas*, but somewhere down the core you would see pollen from *Dryas* or even pieces of *Dryas* plants. Going farther down along your core, *Dryas* would disappear, and reappear, and disappear, and reappear. Counting back, the appearances of *Dryas* would be the Younger Dryas, the Older Dryas, and the Oldest Dryas. Below that, you might find deposits from the glacier that hollowed out the bog. Radiocarbon dating would allow you to say that the Younger Dryas ended about

11,500 years ago, with the others spread over the few millennia before that.

You would infer that the region around your bog had alternated between colder and warmer conditions. The fossils in your bog core would show evidence of younger climate changes, including the recent Little Ice Age cool event from about 500 to 100 years ago, and a larger but rather short-lived cooling about 8,200 years ago, but that the end of the Younger Dryas was the last of the huge changes. But how huge were the changes, and how rapidly did they happen? The answers—really huge and really fast—are seen most clearly in the ice cores from Greenland.

Over the Cliff

Standing in the science trench in Greenland, I measured how thick the annual layers were in the GISP2 core across the end of the Younger Dryas. I found that, going back in time, many thick layers were followed by one slightly thinner layer, one scarcely more than half as thick, another slightly thinner than that, and then a lot of similarly thin ones grouped around a spike of thicker ones. This is most directly interpreted as a twofold change in three years, with most of that change in one year, and with a "flicker" when the climate bounced up and down.

In discussing snow accumulation, it is wise to remember that a year may look anomalous because the drill happened to hit a snowdrift. Occasionally we misidentify a year, which would give an erroneous snow accumulation. The dating down to the end of the Younger Dryas in our ice core may have errors of one year in one hundred, and in older ice we may have errors of a few years in one hundred. So I cannot insist that the climate changed in one year, but it certainly looks that way. And I can insist that the change was fast—not over a century, not even over a human generation, but maybe over a congressional term or even less.

Other records show similar dramatic shifts. Concentrations in the ice of almost all of the windblown contaminants dropped as the snowfall rose, as measured by Paul Mayewski, Greg Zielinski, and others. Correcting for the effect of changes in snowfall shows that the atmosphere of the Younger Dryas had about three times more sea salt, four times more tiny dust particles, and up to seven times more large dust particles than during the first warm millennium that followed, which in turn was about twice as "dirty" as the typical millennia of the current warm period. A careful analysis of the different records by Kendrick Taylor shows that the Younger Dryas ended in three steps, each about five years long or less and spread out over forty years, but that most of the change happened in the middle step. Add in some "flickering" behavior during the end of the Younger Dryas, and it was a weird time indeed.

The stable isotopes of the ice shifted rapidly with the other indicators. The most direct interpretation of the stable-isotopic record based on borehole temperatures is that the surface of Greenland warmed by about 15°F in a decade or less. And this phenomenal change is supported by a completely independent thermometer.

Air contains molecules of different types, and isotopes of those molecules of different weights. Normally, the wind keeps all of these mixed together. If wind mixing somehow stopped, the gases would tend to separate. Heavy ones would settle to the bottom of the atmosphere near Earth's surface, with lighter ones above. Old snow, called firn, must be buried 200 feet, more or less, before it is squeezed into ice. The wind doesn't blow in this firn layer, but the air can still diffuse slowly through the spaces between the ice particles, before the spaces are squeezed closed to make bubbles. The molecules of gases in the firn thus can separate under gravity. The air trapped in bubbles at the bottom of the firn is not identical to that above—gravity has slightly enriched the deeper air in the heavier molecules. The amount of this small enrichment agrees beautifully with expectations based on the known physics of gases. (This enrichment is so small, though, that it does not affect our conclusions about the history of atmo-spheric greenhouse gases.)

There is another way to separate gases by weight. If a strong temperature difference is applied across gases that are not mixed by the wind, the heavy molecules tend to move to the cold end, and the light molecules to the warm end. This was one of several ideas that physicists tested during World War II to obtain a large amount of one isotope of uranium to use for making atomic bombs.

Suppose there was a sudden warming at the surface of the ice sheet at the end of the Younger Dryas. The tempera-ture difference between the surface and 200 feet down, where the bubbles are trapped, would cause the gases to separate by weight, with heavier gases going down through the firn to be trapped in bubbles. The warm surface also would warm the snow and firn beneath it, and eventually this would warm the ice at the bubble-trapping depth, eliminating the tempera-ture difference across the firn and thus eliminating the effects of this temperature difference on gases. But the heat that moves down with the gases must warm the ice it passes as well as warming the air, so the heat flow is slower than the gas changes. The heat from the surface warming would take a century or more to move 200 feet down through the firn to where bubbles are formed, but the gases will separate by weight in response to the temperature difference in about ten years or fewer. For a few decades after the warming, bubbles forming 200 feet down would have gases that are anomalously heavy because of the temperature difference and because of gravity. As the deeper layers warmed, the temperature effect would disappear, leaving only the gravity effect.

Jeff Severinghaus, a geochemist now at the Scripps Insti-tution of Oceanography, analyzed gases in the GISP2 core and discovered the anomalously heavy gases from the warm-ing that ended the Younger Dryas. Analyzing how big the anomaly was, and how rapidly it developed, he found that the warming was about what was expected from the isotopes and borehole temperatures—about 15°F—and that the warming happened in about a decade or possibly even faster.

Jeff made an even bigger discovery, which showed that much of the world's climate is coupled to that of Greenland. A change in the air composition over Greenland will begin to appear in bubbles trapped about 200 feet down in about ten years, but the ice trapping these bubbles typically is centuries old. The bubble-trapping depth, and the age of the ice there, can vary a little depending on how warm or cold the firn is. Just as it is easier to make snowballs out of warm snow than out of really cold powder, snow turns to ice more rapidly in warmer places. The bubble-trapping depth also varies depending on how much it snows—where snow accumulates more rapidly, snow is buried to a greater depth before it has time to be squeezed all the way into ice. We have good models of how changes in snowfall and temperature affect the speed at which snow changes to ice, but these models are not perfect, so there always has been a little uncertainty in figuring out just which gas samples and ice samples are of the same age.

Jeff Severinghaus brilliantly bypassed this uncertainty. He could identify the gas that was trapped in bubbles a decade after the surface warmed suddenly, because the air in those bubbles had the slightly anomalous isotopic composition produced by that warming. And he could ask whether the atmospheric composition changed at the same time as that warming.

Jeff found that the amount of methane in the atmosphere rose just after Greenland warmed. One sample contained a low level of methane, but the signature that the surface had just warmed in Greenland. The next sample, about thirty years younger, had a significantly higher level of methane. (Other people want to analyze ice for other things, so Jeff currently cannot use all the ice to get better time resolution until we get more cores.) Methane continued to rise for a century or so, increasing by about 50 percent.

Methane is very important in our story. Today, human activities produce a lot of the methane in the air. Before there were so many of us, methane was primarily swamp gas, generated naturally by bacteria in the wetlands of the world. Most of the wetlands are either in rainy regions of the tropics and subtropics, or in the tundra and taiga of the far north. Much evidence shows that both high-latitude and low-latitude wetlands must have expanded to explain the methane rise that occurred at the end of the Younger Dryas. During the Younger Dryas, methane was reduced because some of the tropical wetlands had dried out, and some of the high-latitude wetlands had frozen or dried out. After Greenland warmed, the tropical wetlands refilled and the polar ones thawed and filled over the next century or so. But wetlands started to fill or thaw and produce more methane within a few years or decades after Greenland warmed.

The increase in methane after the Younger Dryas would have contributed a little more greenhouse warming. But Greenland warmed abruptly, and then methane rose more slowly (it may take a while to fill or thaw a wetland, and to get the methane-producing bacteria working at full speed), so the warming was not primarily caused by methane. Carbon dioxide didn't change much just after the Younger Dryas— the great reservoir of carbon dioxide in the ocean damps the swings in the atmosphere and causes natural carbon dioxide changes to be slow. So carbon dioxide didn't cause the warming, either. We will need to look at a few other things before we can figure out what did cause the warming at the end of the Younger Dryas.

All Together Now

The really important result from the methane record is that climate conditions changed at the end of the Younger Dryas across a significant fraction of Earth's land surface, including in the tropics, at almost exactly the same time that Greenland experienced its abrupt warming. Dust—probably from China—dropped sharply, sea salt dropped sharply, snowfall rose, and the world's wetlands expanded at the same time, within a period of years to a very few decades.

The warming in Greenland should have increased the snowfall rate there, but the observed change in snowfall was even larger than can be explained by warmer air delivering more moisture. This suggests that the warming caused Greenland to become stormier. Yet the drop in sea salt and dust shows that typical windiness had dropped. A possible explanation is that the storm track—the favored path followed by storms—shifted north toward Greenland as the warming occurred. During the cold Younger Dryas, the wet southern sides of storms delivered moisture to southern Europe while their dry northern sides blew dust over Greenland. With the warming, the storms weakened but shifted their wet sides toward Greenland. The storms typically follow the front where warmer and colder air meet, so this suggests that the warm-air/cold-air boundary moved closer to Greenland as the Younger Dryas ended. Hence, the atmospheric circulation must have changed over large regions.

We have long known that the Younger Dryas occurred in other places—it was discovered by the early researchers studying European bogs, after all. But remember that most of our dating includes small errors. If we are interested in a change that happened over just a few years, it is very difficult to tell whether the change happened first in Africa, Europe, or Greenland, or whether all changed at the same time, because the uncertainties in the ages of the events in Africa, Europe, and Greenland may be a century or more. Because the ice cores collect records of local conditions (snowfall, temperature), regional conditions (windblown dust and sea salt from beyond Greenland), and hemispheric to global conditions (the methane concentration of the air), the ice cores allow us to learn whether the end of the Younger Dryas affected large areas at almost exactly the same time. It did.

We now have strong evidence that during the Younger Dryas, the North Pacific cooled, glaciers advanced from high peaks in New Zealand and the Andes, and lakes in Africa shrank as the Sahara Desert spread into formerly fertile regions. For more than a millennium, Earth was locked in generally cold, dry, windy conditions, although a few places were wetter because storm tracks had shifted over them. Where we have records with high time resolution, the Younger Dryas appears to have ended abruptly. For example, an ocean-sediment core from near the coast of Venezuela records the strength of the trade winds. These winds mix the water, bringing nutrients to the surface to fertilize plankton, which make shells that color the sediment white. The stronger the wind, the whiter the sediment. A graph showing the changes in color of the Venezuelan sediments looks almost identical to a graph showing temperature in Greenland, with cold Greenland corresponding to windy Venezuela. The end of the Younger Dryas took less than a decade in the Venezuelan record, as shown in Figure 12.1.

FIGURE 12.1

The history of windiness offshore Venezuela from Konrad Hughen and colleagues, and the history of snow accumulation in central Greenland from me and many others, as cited in the Sources and Related Information. Ages were determined independently, and have small errors. Lines show probable correlations between the records. The end of the Younger Dryas cold event in Greenland snow accumulation occurred in one to three years; the change in windiness off Venezuela occurred in ten years or fewer.

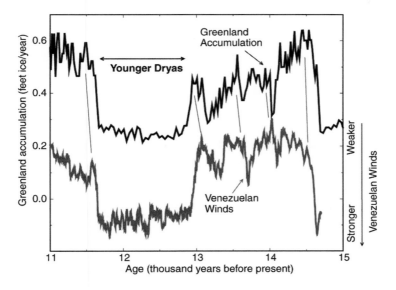

With one partial exception about 8,200 years ago, the ice-core records show no similarly large, abrupt changes in snow-fall, temperature, dust, or methane since the end of the Younger Dryas. The millennia over which agriculture and industry rose have been calm and constant by comparison. True, climate changes have contributed to the rise and fall of empires, lured the Vikings to Greenland and then driven them out, and otherwise affected human lives. But these changes that have affected historical humans appear as slow one-degree shifts in the ice-core records, not as abrupt ten-degree jumps. The large effects that small climate changes have had on humans, and the unequivocal records of much larger climate changes, are enough to make some people think deeply, and even to make them a little nervous.

Turning More Worms

The Younger Dryas encompassed 1,300 years of cold, dry, and windy conditions across much of the world. Warmth returned after the Younger Dryas with a bang, not a whimper. But when we back up and look at 100,000 years of Greenland ice-core record, the Younger Dryas is just the "same old stuff." A similar or larger jump occurred about 15,000 years ago. Then the climate cooled slowly, cooled in a jump, warmed in a jump, and repeated as it reeled into the Younger Dryas. Dozens of rapid changes litter the record of the last 100,000 years as indicated in Figure 12.2. If you can possibly imagine the spectacle of some really stupid person (or, better, a mannequin) bungee-jumping off the side of a moving roller coaster car, you begin to picture the climate—the roller coaster rides the orbital rails of the ice ages, with the bungee-jumping maniac bouncing up and down past it. At least some of the abrupt changes show "flickering" behavior, with the climate jumping back and forth between two states for a few years before settling down into one of the states.

Over the last 100,000 years, there were only two vaguely stable periods of climate. The first was when the ice sheets were biggest and the world coldest. Then, Greenland stayed

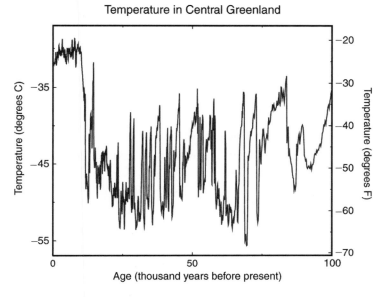

FIGURE 12.2
The history of temperature in central Greenland over the last
hundred thousand years, from ice-isotopic values calibrated against
borehole temperatures, using data from the 1997 paper by Cuffey
and Clow. The prominent Younger Dryas cold event is now seen to
be "business as usual," with similar events having dominated the
record. Jumps have been smaller than usual around 20,000 years
ago, when most of the world was in the coldest part of the ice age,
and during the most recent few millennia, when most of the world
was warm. Fahrenheit temperatures are shown on the right, and
Celsius on the left.

cold, the air stayed dusty, and methane levels stayed low for a
few thousand years. Just as a bungee jumper on a roller
coaster car can't go farther down than the ground, perhaps
the climate was just as cold as it could get. And it was very
cold. Central Greenland was more than 40°F colder than today, on average. If my home in central Pennsylvania were to
cool that much today, the kite-flying days of May would turn
into the frosty mornings of December! The second vaguely
stable climate is the period we are living in now. Over the last
few thousand years, Greenland has stayed warm, not much

dust has blown through the air, global wetlands have stayed wet and continued to produce methane, and humans have figured out how to grow crops and build cities in the comfortable climate we enjoy.

But for most of the last 100,000 years, a crazily jumping climate has been the rule, not the exception. Slow cooling has been followed by abrupt cooling, centuries of cold, and then abrupt warming, with the abrupt warmings generally about 1,500 years apart, although with much variability. At the abrupt jumps, the climate often flickered between warm and cold for a few years at a time before settling down. One can almost imagine a three-year-old who has just discovered a light switch, flicking it back and forth, losing interest for a while, and then returning to play with it again.

The abrupt warmings recorded in Greenland ice were first described by the Danish geochemist Willi Dansgaard and coworkers in the late 1960s and early 1970s, based on analyses of the Camp Century ice core from northwestern Greenland. However, the records of these events occurred in ice from very close to the bed, where flow thinning made the history difficult to read, and where flow over bedrock bumps may have distorted the records. In the mid-1980s, Willi Dansgaard, his Swiss colleague Hans Oeschger, and other collaborators described essentially the same history from the Dye 3 core from south Greenland. Two records are much harder to ignore than one, so the scientific community began to pay attention to what we now call Dansgaard-Oeschger events or cycles in honor of these pioneers. However, Dye 3 also had the really interesting ice very close to the bed where ice flow had made the records difficult to read.

Our central Greenland ice cores were drilled in the very best part of Greenland for finding the record of the Dansgaard-Oeschger cycles far above the bed and in wonderful condition. Drilling two such cores allowed comparisons to dispel any doubts about the quality of the records. So we now can describe these rapid climate changes with confidence over the last 110,000 years or so.

But new knowledge always brings new questions. Is the

last ice age unique in having Dansgaard-Oeschger cycles, or
have such cycles occurred before? Are all warm periods stable? How long do stable periods last? Is our spell of good
weather almost over?

We cannot answer most of these questions in central
Greenland. The high snowfall rate that allowed us to recognize and count the last 110,000 years of history has caused the
ice sheet to be large and steep and so flow rapidly, greatly
thinning and deforming older ice. The ice in both central
Greenland cores older than about 110,000 years has been
folded or otherwise disturbed, and the climate record is no
longer continuous. As I write this, a primarily European consortium is hard at work on a NorthGRIP project to obtain a
core in an area with slightly lower accumulation north of the
central Greenland cores, hoping to find an intact history
through the previous warm period. Antarctic cores from
places with even less snowfall provide insights to longer
times. But for really long records, and to learn about ocean
processes not recorded in ice cores, we need to go back to
the sea floor. Although worms have stirred up much of the
mud in the ocean, some of the sediment is not badly worm-
burrowed. Careful searches by scientists such as geologist
Gerard Bond of the Lamont-Doherty Earth Observatory of Columbia University have found such cores.

Gerard Bond measured how cold the north Atlantic was
in the past by comparing the abundances of shells of cold-
loving and warm-loving creatures in his cores. He also counted the rocks in his cores. Dust blows across the oceans, and
occasionally sea ice or driftwood will move rocks, but the
only known way to get lots of large rocks to the middle of the
Atlantic is to have them carried out in the bottoms of icebergs
and then released as the bergs melt. Reassuringly, Gerard's
records were almost identical to those from the Greenland ice
cores. The surface of the north Atlantic was cold whenever
Greenland was cold, and the north Atlantic surface was warm
when Greenland was warm. During the general cooling to the
most recent ice age and the warming from it, the north Atlantic experienced numerous abrupt coolings and warmings,

with each abrupt warming typically 1,500 years after the previous one, although with much variability.

In comparing ocean and ice records, Gerard Bond noticed another cycle. On land and in the ocean, after an especially large warming, the next one 1,500 years later wouldn't get quite as hot, with the next another 1,500 years later and the next 1,500 years after that a bit cooler still. After three, four, or five of these progressively cooler Dansgaard-Oeschger cycles, there would be another really big warming. And the coldest time just before that big warming was also a time when the north Atlantic was inundated by icebergs. This pattern is not reproduced perfectly each time, and it is complicated by the 19,000-year, 23,000-year and 41,000-year orbital cycles that sometimes combined to cause warming, opposing the cooling trend of the successive Dansgaard-Oeschger cycles. However, most workers recognize this pattern as an important part of the climate system.

Purgery and Icy Armadas

The staggering cycle of a few thousand years of cooling leading to an iceberg inundation, and then a few years or decades of spectacular warming, is now known as the Bond cycle. It took many workers a few frantic years to explain this bizarre pattern. We will try to follow them.

Most sediment in the north Atlantic contains at least a little iceberg-rafted debris, and sediment from colder times has more rocks carried by bergs. However, a few anomalous layers are almost entirely iceberg-rafted debris. Six such prominent layers were recognized in sediments from the last 100,000 years by the German researcher Hartmut Heinrich (see Figure 12.3).

These Heinrich layers occur across the north Atlantic. Each is only a fraction of an inch thick on the east side of the ocean, but the debris layers thicken to the north and west toward Hudson Bay, to more than a foot thick just outside the mouth of the bay. Rocks from many lands around the north Atlantic can be found in the thin edges of the Heinrich layers,

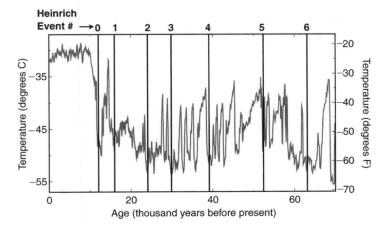

FIGURE 12.3

The history of temperature in central Greenland as in Figure 12.2, with the times of Heinrich events (numbered 1 to 6, plus the Younger Dryas, which is often referred to as Heinrich event 0 because numbers 1 through 6 had already been used and no one wanted to renumber them), as identified by Gerard Bond. Gerard also identified the cycle of progressive cooling of successive cold jumps, followed by a Heinrich event and an especially large warming, which we call the Bond cycle.

but the thick parts of the layers are dominated by rocks of types that are common in and around Hudson Bay, but rare elsewhere. The layers were deposited very rapidly, at a time when the ocean was exceptionally cold.

Gerard Bond showed that an especially warm time occurred just after each Heinrich layer, with temperatures then staggering colder until the next Heinrich layer. It fell to Doug MacAyeal of the University of Chicago to explain the massive, rapid outpourings of bergs and debris that produced the Heinrich layers and, indirectly, the Bond cycle (Figure 12.4).

When Earth's orbit caused northern summers to receive little sunshine, snowfall began to grow an ice sheet on the Canadian highlands around Hudson Bay. These regions have permafrost today—even in a warm climate, the yearly average temperature is below freezing and a layer in the ground remains frozen all year, although the hot summers allow

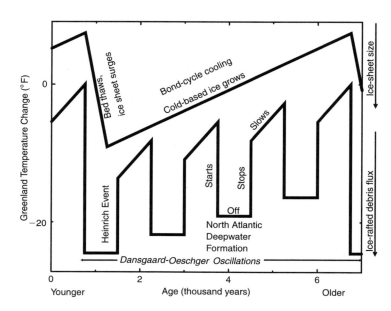

FIGURE 12.4

An idealized history of temperature and ice sheet changes in the
north Atlantic region during a Bond cycle. Successive Dansgaard-
Oeschger oscillations, caused by the turning on and off of the far
northern sinking of waters in the north Atlantic, become
progressively cooler as the cold-based ice sheet grows in Hudson
Bay. Then the base of that ice thaws, and a Heinrich event surge
occurs, dumping large numbers of icebergs containing rock debris
into the north Atlantic. When the surge ends, the ice sheet freezes to
its bed, while the ocean circulation resumes and causes an especially
large warming away from the ice sheet.

the snow and the very top layer of the ground to melt, and
very deep ground is thawed by Earth's heat. Cooler summers
would have allowed the snow to survive and grow a glacier,
but initially that ice would have frozen to the cold permafrost
beneath. As the ice sheet thickened, feasting on the snowfall
of the ice age, the cold ice would have slowly spread from
the highlands to fill Hudson Bay. The flowing ice would have
carried its cold with it, just as you can bring cold water from
underground to your shower on a hot day, so the ice that

filled Hudson Bay would have frozen to the mud and rocks beneath the bay.

Piling ice on top of the land tends to trap Earth's heat, however, just as a pile of blankets keeps heat around you in bed rather than letting that heat escape to the air. More ice acts like more blankets. If you are covered too deeply, you may feel like you're melting—if it is covered too deeply, the bottom ice really is!

The ice in Hudson Bay would have built up through several Dansgaard-Oeschger cycles, growing rapidly during the cold times and slowly during the warm times, until the ice was thick enough to trap enough of Earth's heat to thaw the bed. Then, the mud, stones, and water would have allowed the ice to "skate" very rapidly over deeper bedrock, dumping icebergs and their mud and stones into the north Atlantic. In just a few centuries, Hudson Bay would have "purged" itself of the ice it had "binged" over millennia. The mud and stones from the icebergs explain the Heinrich layers, and the icebergs would have chilled the surface of the ocean. Eventually, as the ice thinned, its cold surface layers would have moved near the bed and frozen to it again. Then, the ice motion would have nearly stopped, and the ice sheet would have begun building up for the next cycle.

This cycling would end only when orbital or other changes made the climate too hot to allow the ice to grow again. The Heinrich events early in the ice age were spaced further apart in time than those near the peak of the ice age, probably because it took longer for the ice to thicken when Earth's orbit was less favorable for ice growth.

When the ice sheet was growing in Hudson Bay, the great bulk of the ice would have pushed the winds aside and cooled its surroundings with the cold air that flowed down its flanks. After a purge, the smaller amount of ice remaining in Hudson Bay would have allowed warmer conditions nearby. The slow growth and rapid shrinkage of the ice in Hudson Bay thus may explain the slow cooling and rapid warming of the Bond cycle, with the Heinrich event at the coldest point.

The ice in Hudson Bay was enough to flood the north Atlantic with icebergs, but was only a part of the vast ice sheet that covered almost all of Canada and the northern tier of the United States. The lower elevation of the bed of Hudson Bay apparently caused its ice to be thicker and warmer and so to melt at the bottom sooner than other ice did. The Hudson Bay ice thus oscillated every few thousand years, while the bulk of the North American ice sheet changed on the slower orbital time scale of tens of thousands of years.

Yo-Yos

By now, it may be getting hard to keep score, so let's recap. During the last million years, ice grew for about 90,000 years as we sank into the global cold of an ice age, with smaller wiggles spaced about 19,000, 23,000, and 41,000 years apart. Ice then shrank for 10,000 years as we warmed into an interglacial period. This pattern repeated, over and over, in response to changing sunshine linked to features of Earth's orbit. The climate was most nearly stable (although still quite variable) during the coldest and warmest times. But during the coolings into the ice ages and warmings from ice ages over the last 100,000 years (and, as we shall see, during much of at least the last million years), the climate in the north Atlantic jumped between several cold centuries and several warm ones. Through a few of these Dansgaard-Oeschger cycles, each warm event was a little cooler than the one before. Then the ice sheet in Hudson Bay suddenly dumped a great mass of icebergs into the north Atlantic in a Heinrich event. The next warming was especially large, and then the Bond cycle of progressively cooler events resumed. You might think of a roller coaster riding the orbital rails, with Heinrich-Bond bungee-jumping off the roller coaster while playing with a Dansgaard-Oeschger yo-yo.

The most recent orbital cycle reached the coldest part of the ice age about 20,000 years ago, followed by a 10,000-year warming trend to reach our current mild, stable climate. The warming trend was punctuated by several abrupt warmings

and coolings. The climate cooled into the Younger Dryas—
the youngest of the large changes—about 12,800 years ago,
and warmed again about 11,500 years ago. Because the
Younger Dryas occurred after the world had begun warming
from the depths of the most recent ice age, the icebergs of
Younger Dryas time melted faster and did not carry rocks as
far as some older icebergs did. But in many places close to
Hudson Bay, the Younger Dryas looks like a seventh Heinrich
event.

Much effort has gone into tracking the footprint of the
Younger Dryas around the globe, with considerable success
as described above. The older Dansgaard-Oeschger and
Heinrich-Bond signals have been harder to follow. Many records of the Younger Dryas have come from lakes and bogs
that were hollowed out by glaciers that began melting only a
little before the Younger Dryas, so these lakes and bogs don't
have older records. Other records become harder to date as
they get older, so correlations are harder to achieve. Finally,
however, the picture is beginning to come clear.

The available data now indicate that the cold times of
both Dansgaard-Oeschger and Heinrich-Bond oscillations in
Greenland and around the north Atlantic also were cold, dry,
and windy in broad regions extending into subtropical Africa
and Asia, and across Europe and North America. Much of the
world probably became drier because the cooling reduced
water vapor, although certain regions became especially dry
as storm tracks shifted away from them to areas that became
somewhat wetter. The overall drying allowed more cooling
because of the greenhouse gas effects of water vapor.

The Heinrich events, including the Younger Dryas, left a
bigger "footprint" than the non-Heinrich Dansgaard-Oeschger
oscillations, stamping a Heinrich signature on the entire
globe, rather than primarily in one hemisphere. But in a truly
bizarre twist, the Heinrich events were warm, not cold, in the
far south Atlantic Ocean and in parts of the Indian Ocean and
the Antarctic downwind of the south Atlantic, whereas the
Heinrich events were cold in at least one site in Antarctica
and in most of the rest of the world. Solving this puzzle will

tell us much about the causes of these events, which will be discussed in the following chapters.

However, we should make two more points before we consider the causes. First, work is progressing on the monumental task of locating and studying very long ocean sediment records that have not been badly stirred by burrowing worms. This requires a huge effort to take many tiny samples and analyze them in many different ways. The upper parts of those records show the familiar yo-yo/bungee-jumping/roller coaster climate from the ice cores. And the deeper, older parts of those records show the same thing. Over most of the last million years, and possibly over much longer times, the constant has been change. And the current period of stable climate is among the longest on record.

Equally puzzling, the current warm time is not perfectly stable, by any means. After all, climate changes have chased the Vikings out of Greenland, dried up the "land of milk and honey," and otherwise perturbed humans throughout our written history. Some of these more recent changes have been linked to subtle coolings and warmings of the north Atlantic. Surprisingly, although these more recent north Atlantic changes are much smaller and slower than the great jumps of the Dansgaard-Oeschger cycles, these more recent changes have the same approximately 1,500-year spacing. This almost certainly provides important information on why the events occur, so we will come back to it later.

For now, though, we have more than enough confusing data. When other climate historians and I were faced with these startling results, we realized that we had to learn more about the climate system so that we could make sense of the stories that the ice and the other sediments were telling. You can join us in this effort in the next part.

IV

Why abrupt

changes

happened to

Earth's climate

in the past

13

The Greenland ice cores and other records show that climate changes large enough and rapid enough to scare civilized peoples have occurred repeatedly in the past, and that our civilization has risen during an anomalously stable time. We would like to understand the climate jumps to learn whether they might happen again, and whether changes in human behavior can make climate jumps less likely.

The jumps have occurred in Earth's wildly complex, linked, feedback-dominated climate system in which atmosphere, oceans, ice, land surface, and living things interact with each other and with the solar system to drive weather forecasters and climate scientists to distraction. In this chapter, we will take a quick tour of how Earth's climate works. Then, in the next chapters, we will use this knowledge to see what might have caused the abrupt climate changes.

Briefly, the planet receives much sunshine at the equator, and little sunshine at the poles. This imbalance drives currents in the air and the ocean, which take the extra heat from the equator to the poles. Because the earth rotates under them, the currents tend to go in circles rather than going straight to the poles, leaving the poles colder than they would be without rotation. The narrow north Atlantic allows ocean currents to "lean on" the continents and flow north rather than turning, making the north Atlantic unusually warm. However, the hot currents in the north Atlantic could be stopped if a little more fresh water were added there, with grave consequences.

Balancing the Books

Earth gets most of its energy from space. The sunshine at the top of the atmosphere supplies about as much energy as three or four bright light bulbs for each desktop-sized area on Earth (about 340 watts per square meter, or 290 watts per square yard). The geothermal heat supplied from deep inside Earth is almost 10,000 times smaller (only about 0.05 watts per square meter), requiring a whole football field to give enough energy to run three light bulbs. Thus, although geothermal heat can be trapped and melt ice under a two-mile-thick pile, Earth's heat really doesn't have much direct effect on the atmosphere.

About 30 percent of the sunshine reaching Earth is reflected right back to space, from clouds or snow or desert sands, and does not heat the planet at all. The rest of the sunshine heats Earth.

All things radiate energy all the time. Warming an object causes it to radiate more energy, and at shorter wavelengths. If you've watched an electric stove, you know that you begin to see the glow of a burner as it warms, going from longer-wavelength deep red to shorter-wavelength orange-red. If the burner warmed still more, it would glow in the shorter wavelengths of yellow, and then white. If your eyes could see in infrared (longer wavelengths), you would know that the burner glows a little at long wavelengths even when it is at room temperature or colder, as does everything else.

Earth's temperature is whatever is required to send back to space the same amount of energy that the planet absorbs. If less energy is sent back than is received, the planet warms, "glowing" more brightly and sending more back until a new balance is reached. If less energy then were to reach Earth, Earth would be sending back more than it received, and Earth would cool and glow less brightly until balance was again reached. This is a very important negative or stabilizing feedback, and helps moderate our climate.

The energy Earth sends back to space is mostly at long wavelengths that humans cannot see, whereas the energy from the sun is mostly at short wavelengths that humans can see, but the total energy sent back is almost exactly equal to the total received from the sun. This is quite similar to an automobile factory that receives little items (wheels and bolts and radio antennas and such) and ships out big products (cars), but still must get rid of as much material as it receives so it doesn't overflow.

If the sun were to get brighter, or if Earth were to move closer to the sun, we would receive more sunshine and would warm up. Earth's orbit does have important wiggles that move sunshine from north to south or summer to winter, but these do not have much effect on the total amount of sunshine we receive. The sun's output varies a little, but probably not much over centuries to millennia. The amount of sunshine reaching the top of our atmosphere thus does not seem to change much over times that really matter to human societies, and the only ways to change the average temperature of the whole Earth are to change either Earth's reflectivity or the ease with which Earth's radiation escapes.

How well sunshine is reflected back to space is called the *albedo*, and can range from 0 percent (none reflected, all absorbed, hot Earth) to 100 percent (all reflected, none absorbed, cold Earth). If Earth reflected less energy, more would be absorbed and Earth would warm up until as much energy was radiated as was absorbed. Clouds, especially the thick ones close to the surface, reflect a lot of sunlight—if you've looked down on a cloud from an airplane on a sunny day, the cloud looked bright because it was reflecting the sunlight back up toward space. A reduction in highly reflective low, thick clouds would warm Earth. Snow and ice work like clouds to reflect sunlight, so if some warming caused snow and ice to melt, Earth would absorb more sunshine and warm up a little more in an ice-albedo positive feedback.

The other way to warm Earth is to block some of the outgoing longwave radiation. If we suddenly erected such a blockade, Earth would be receiving more radiation than was going back to space, and the surface would begin to warm. This warming would increase the outgoing energy trying to "run the blockade" until a new balance was reached, with most of the change happening in a few days.

If you are cold in bed, you might warm up by putting on a blanket to block some of the heat leaving you by radiation and in other ways. Earth's "blanket" is the atmosphere. Water vapor is especially important. Shortwave sunlight passes through water vapor easily, but the longwave radiation that Earth sends back is partially blocked by the water molecules in the air, warming the air and the surface of Earth beneath. Water and some other molecules in the atmosphere absorb those wavelengths that have just enough energy to cause the molecules to wiggle, vibrate, or spin, while wavelengths with higher or lower energy pass through. In a vaguely analogous way, the tire on your car will pass over tiny bumps (sand grains) and big bumps (hills), but be "absorbed" by in-between ones (potholes).

The tendency of some gases in the atmosphere to catch outgoing radiation, "blanketing" Earth and warming it, is called the "greenhouse effect"—a greenhouse lets sunlight in more easily than it lets other energy out, causing the greenhouse to be warmer than its surroundings. Carbon dioxide, methane, and other gases contribute to the greenhouse effect, although water vapor is most important. Without the greenhouse effect, much of Earth's surface would be permanently frozen, and we would be unhappy (if we were here at all). Natural changes have occurred in the planet's albedo and greenhouse effect, and natural and human-caused ones are occurring and will continue to occur.

Putting the Right Spin on It

The greenhouse effect and Earth's albedo help control our temperature, but simple geometry is the most important factor in our weather. The curvature of the planet causes sunlight to hit Earth with a "glancing blow" at the poles, spreading a given amount of sunlight over a larger area than at the equator as shown in Figure 13.1. Space surrounds Earth on all sides, so the poles and the equator can radiate heat to space easily. Because they receive less energy but radiate as well, the poles are colder than the equator. (The equator also is closer to the sun than the poles, but by only 0.004 percent of the distance from the sun, which is not enough to matter.)

If we didn't have an atmosphere and an ocean, the equator would be too hot for humans, and the midlatitudes and the poles too cold. Fortunately, we have air and water, which transfer extra heat from the equator to the poles. Air at the surface in the tropics is heated, expands, rises, and then flows

FIGURE 13.1

The curvature of Earth causes a given amount of sunlight to be spread over a much larger area near the poles than near the equator, as shown. This causes the poles to be cold and the equator to be hot.

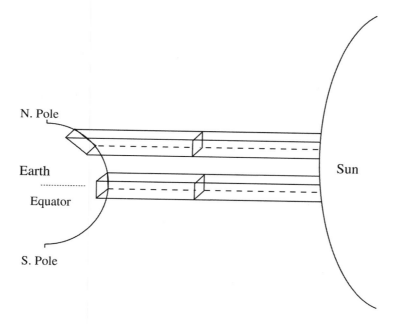

toward the poles. The air cools on the way, in part by radiating heat to space and in part by heating the land or water beneath and letting them eventually radiate the heat to space.

The climate could be this simple—hot air rises at the equator, flows away and warms colder places as it does, sinks at the poles, and returns to the equator to warm up again. But because Earth rotates on its axis, poleward-moving air spins over the land in an inefficient system that keeps a large temperature difference between equator and poles.

If you were to walk and swim around Earth at the equator, you would travel 25,000 miles to come back to your starting point. But you can walk around the Arctic Circle, or the Antarctic Circle, in only 10,000 miles. And I've walked around the South Pole in 3 steps. If you stand still at the equator instead of walking, Earth's rotation will carry you 25,000 miles eastward in a day, bringing you in sight of the sun (sunrise), under the sun (noon), and on to sunset and another sunrise at more than 1,000 miles per hour. The same cycle will be achieved in 10,000 miles in a day or 400 miles per hour at the Arctic or Antarctic circles, and exactly at either pole you will spin about your spine without going anywhere.

If you stand on the equator, however, you aren't buffeted by thousand-mile-per-hour winds. The air interacts with the trees, mountains, and waves on the surface, and so moves about as rapidly as Earth's surface beneath it. It is a curious property of liquids and gases, including the wind, that when they are very close to a moving body they tend to move with the surface. No surface is perfectly smooth, and the roughness drags a fluid along (as do chemical interactions between air and surface). Thus, you can drive your car down the freeway at sixty miles per hour and the dust and bugs won't blow off the windshield, despite the fact that a sixty-mile-per-hour wind is more than strong enough to carry dust and bugs—the wind drops to zero very close to the windshield. Similarly, a hundred-mile-an-hour baseball pitcher can throw a dirty ball, and the dust will still be on the ball when it hits the catcher's mitt. So air near the equator is moved eastward by Earth's spin at almost 1,000 miles per hour, but air near the Arctic

Circle is moved eastward at only about 400 miles per hour, and air very near the North Pole is not moved by Earth's spin.

Earth's rotation is the source of the Coriolis effect, which causes the winds to curve over the ground. Suppose you watch a parcel of air that starts at the equator, moving eastward with the ground beneath it at 1,000 miles per hour. If this air parcel begins to move northward, it will pass over land that is spinning at 900 miles per hour, then 800, then 700. But air very far above the trees, mountains, and waves will take a while to be slowed down much by friction with them, and so will continue to move eastward at about 1,000 miles per hour. If you were to stand on the equator and watch the air going away from you toward the United States, the air would move to the east faster than the land beneath it, and so would seem to curve to the right and "get ahead" of the land. An air mass starting out at the Arctic Circle and moving away from you toward the south would soon find itself over land that is moving east faster than the air, and so the air would seem to lag behind the land, again seeming to turn to the right. All air masses moving away from you appear to turn to the right in the northern hemisphere, and to the left in the southern hemisphere.

The Coriolis effect has some interesting implications. Winds tend to blow away from places with high pressure, and in the northern hemisphere the turning of these winds will make a clockwise circle when viewed from above. Winds moving into a low-pressure center, such as a hurricane, will turn to give a counterclockwise flow in the northern hemisphere when viewed from above. Opposite turns in the southern hemisphere will give clockwise low-pressure storms and counterclockwise flow away from high-pressure areas when viewed from above. (Contrary to the stories you may have heard, the difference in rotation speed across your bathtub or toilet bowl is so tiny that the direction the water spins going into the low-pressure system of the drain has more to do with the design of the bowl than with the Coriolis effect. I have observed fluids rotating both clockwise and counterclockwise going down different drains in either hemisphere.)

Air rises in the tropics, cools as it heads for the poles, but also begins to turn so it cannot reach the poles directly. Instead, the air ends up sinking in the "doldrums" or "horse latitudes" only about a third of the way to the poles, and starts flowing back toward the equator, again turning as it does and giving us the trade winds, as seen in Figure 13.2. If you like

FIGURE 13.2

A greatly simplified diagram of Earth's atmospheric circulation. Surface winds are shown within the circle outlining Earth, and a vertical profile through these surface winds and their return flows is shown around the outline of Earth. The regions of rising air tend to be wet, and the regions of descending air tend to be dry, as indicated. The curving of the winds over the land surface is caused by Earth's rotation, shown by arrows around the poles.

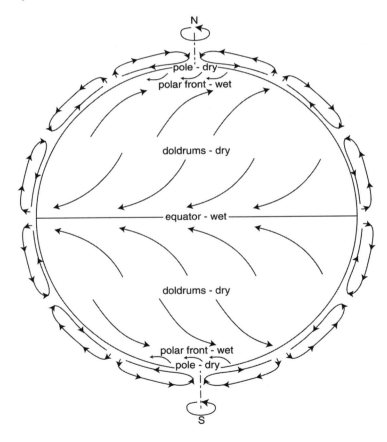

fancy names, the air rises near the equator in the *intertropical convergence zone,* and the entire loop from the convergence zone to the doldrums and back is the *Hadley cell.*

Rising air expands as it is freed from the pressure of more air above, and sinking air is compressed as more air flows in above. Air cools as it expands and warms as it is compressed, as shown by the coolness of air rushing out of a puncture in a high-pressure bicycle tire and the warmth of the bicycle pump after you use it to refill the patched tire. The cooling of rising air causes moisture to condense and make rain clouds, so the rain forests of central Africa and Brazil sit close to the equator under the intertropical convergence zone. Rain forests also develop where wind blowing against mountains is forced to rise, as along the west coasts of the United States and the islands of New Zealand. The warming of sinking air causes evaporation, producing the Sahara and the Kalahari deserts of the doldrums. Deserts also form where air descends after being forced over mountains, as in Death Valley and other regions east of the U.S. mountains, and in the desert east of Tongariro and its neighboring volcanoes in New Zealand.

Far toward the poles, there is another "thermally direct" circulation loop with air rising, moving poleward, and then sinking again in a polar cell. Between the Hadley and the polar cells, across Europe, North America, and parts of Australia and New Zealand, the average airflow is opposite to what one might expect, with sinking motion closer to the equator, rising motion farther away from the equator, poleward motion along the surface, and equatorward motion aloft. But if you think of the different "loops" as gears, you will see that the middle loops must do what they do, turning "backward" to link the "forward-turning" flows near the equator and the poles. The Coriolis turning of the poleward surface flow in the midlatitudes gives the "prevailing westerlies" of the world, causing the west sides of the mountains of New Zealand and the United States to be the wet sides. Fronts occur where poleward- and equatorward-moving air meet, and wiggles on those fronts produce storms that make weather forecasting interesting.

All this complexity ultimately is part of Earth's climate control system, cooling off the equator by shipping some of its excess energy to the poles to warm them up. But the atmosphere moves only about half the heat, with the other half going through the oceans. Ocean circulation is, in some ways, even more complex than that of the atmosphere, because ocean currents can be caused by the wind and by differences in water temperature or salinity. We will see that the interplay between ocean temperature and salinity is at the root of the crazy climate stories recorded in the Greenland ice cores.

Overturning the Ocean

Blow across a coffee cup, and you will see your "wind" drag some of the coffee to the other side of the cup. Your "coffee current" will split against the far side of the cup and come back along the rim, and some of the current may sink. On the near side of the cup, some "upwelling" occurs, with coffee from below coming to the surface to join that coming along the rim.

The trade winds blow across the "coffee cup" of the oceans, from Africa across the Atlantic to Brazil, and from Peru across the Pacific toward Australia. These winds drag water along. This moving water leaves partial "vacuums" or low-pressure zones along the west coasts of Africa and South America, which are filled by water upwelling from below and by surface currents along the coasts.

In addition, the northern-hemisphere and southern-hemisphere trade winds blow toward each other along the equator to feed the rising air of the intertropical convergence zone. The Coriolis effect applies to water as well as to air, and causes water to move to the right of the wind's direction in the northern hemisphere and to the left of the wind's direction in the southern hemisphere. As the trade winds flow westward and toward the equator, they cause currents that flow westward and away from the equator, allowing more water to rise from below to fill the gap. The water near the equator is warmed by the sun more intensely than any other water on Earth, so forcing water out of the tropics is another way that the Earth system transports excess equatorial heat toward the poles. Much of the water moving north forms the great Gulf Stream of the Atlantic and the Kuroshio Current of the Pacific, with the Brazil and other currents flowing to the south, vaguely like the coffee running from your breath along the rim of the coffee cup.

Just as cold air at the poles sinks to the bottom of the atmosphere, cold water at the poles sinks to the bottom of the ocean, filling almost the entire ocean with polar waters and leaving only a thin warm layer on top in the low and middle latitudes. But the ocean has an additional complication—salt. A spoonful of sugar added to your coffee will sink, taking some of the coffee along. If you were crazy enough to add salty water to your coffee, the salty water would sink. The densest waters in an ocean, or a coffee cup, can sink and make currents even if the wind is not blowing, and water can be made denser by making it colder or saltier.

Evaporation leaves salt and takes fresh water from the ocean's surface; that fresh water eventually is returned to the ocean by rainfall, by rivers from continents, and by melting icebergs. Where evaporation exceeds return, the surface water of the ocean becomes saltier. Where return exceeds evaporation, the ocean surface becomes fresher. The changes are small—waves, storms, and currents tend to mix the fresher and saltier parts of the ocean and even out the differences in the same way that your spoon stirs sugar into the coffee—but the differences are real.

Today, the surface water of the Atlantic Ocean is typically saltier than the surface water of the Pacific, Indian, or Arctic oceans. The Atlantic probably is salty because the trade winds blowing westward across the low, narrow land of Central America take water vapor that evaporated in the Atlantic over to fall as rain on the Pacific. The trade winds blowing from the Pacific have to cross the Indian Ocean and Africa to get to the Atlantic, and so have trouble returning water to the Atlantic.

Whatever the reason, the Atlantic has salty, dense water.

Where salty water cools, near Antarctica and in the far northern Atlantic, it becomes the densest water in the world ocean and so sinks to the bottom. Eventually, the stirring and mixing in the ocean make this deep water slightly warmer and less salty (though still cold and salty), and it rises from the deep ocean and is replaced by colder, saltier water.

The water that cools and sinks near the poles must deposit its heat somewhere. Ultimately the heat is lost to space, but first the water warms the air above it. Most of this water-cooling/air-warming occurs in fall and winter, when the air is colder than the water.

People who live downwind of the Great Lakes of the United States and Canada know that when bitterly cold arctic winds blow over the open water of the lakes, the winds become warmer and wetter, and then dump their newfound moisture and heat on Buffalo, New York, and Erie, Pennsylvania. While the Great Plains are suffering under dry, forty-below weather, the "snow belt" is disappearing under a white blanket only a few degrees below freezing.

A similar effect happens downwind of the north Atlantic. Roses thrive in Britain, at the same latitude where polar bears live near Hudson Bay, because the British winters are made warm and wet by the north Atlantic. Warm and salty ocean currents enter the north Atlantic, lose their heat to the atmosphere, and then sink miles into the deep ocean, making room for more warm water.

When the Great Lakes freeze during extreme winters, Buffalo suddenly becomes cold and dry, crisp and clear. Although ocean water can freeze to make sea ice, and does freeze in many places around the Antarctic and the Arctic, some parts of the north Atlantic have not been known to freeze in the modern climate. We will see soon that the Great Lakes tell us more about past climate than we might care to know.

It often is useful to think of the north Atlantic as the end of the conveyor belt at a supermarket checkout, an analogy championed by the great geochemist Wally Broecker of the Lamont-Doherty Earth Observatory of Columbia University.

Wally, more than any other individual, is responsible for the scientific focus on abrupt climate changes, and for what we know about these changes. He has worked on dating techniques, climate reconstructions, and explaining the events that he has reconstructed and dated. Much of his work has involved ocean sediments, but he has interests in ice cores, pollen, lake sediments, and other recorders. He named, and publicized, the Dansgaard-Oeschger and Heinrich-Bond cycles. Almost everyone who studies climate changes has either worked with Wally, or had a good idea and then discovered that Wally beat them to it, or both.

Wally Broecker is especially noted for generating compelling syntheses of data and models, producing overviews that drive scientists determined to prove that he is either right or wrong. Some refer to these overarching constructs as paradigms, but I find it easier to think of them as "cartoons," distilling vast and vastly confusing reality into a form that we can talk about, test, and use. Some experts on ocean currents grind their teeth when they hear the ocean circulation compared to a conveyor belt, because the ocean is indeed more complicated than shown by this cartoon. But just as a political cartoon can help us understand a complicated scandal in Washington, the "cartoon" of the conveyor, shown in Figure 13.3. is extremely useful in helping us understand the ocean.

Briefly, hot and salty water flows north along the surface of the Atlantic, cools, sinks, and then flows southward in the deep ocean. This water eventually moves into the deep Indian and Pacific oceans, and becomes warmer and fresher as other water is mixed back into it. This deep water then rises to the surface in the Indian and Pacific oceans, and in places around the Antarctic. Some of the Antarctic water sinks again, but much of the water flows back along the surface into the Atlantic, becomes warmer and saltier while flowing north across the equator, and sinks again in the north Atlantic, taking roughly a millennium to complete the great conveyor loop. Heat and moisture are scraped off this conveyor belt into the air over the north Atlantic—groceries for the people of Europe.

Many people wonder why there is so much focus on the north Atlantic. The north Pacific is too fresh for much sinking, so attention is not directed that way. But about as much sinking of colder waters occurs in the Antarctic as in the north Atlantic, so why don't we concentrate on the Antarctic?

The surprising answer is the Drake Passage, that storm-tossed stretch of seasickness between South America and Antarctica. (Having once crossed the Drake in a small research vessel, I report from experience on the seasickness part.) Try to force warm surface waters to flow from the South Atlantic to the Antarctic, and the Coriolis effect causes the water to turn left, making a great loop around the southern continent as the Antarctic circumpolar current. Water moving poleward

FIGURE 13.3

A "cartoon" showing the "conveyor belt" global oceanic circulation championed by Wally Broecker. Although the ocean currents are much more complex, this illustration captures many important aspects of the circulation. Today, warm waters flow northward along the surface of the Atlantic, cool and sink in the far-northern and near-northern parts of the Atlantic, and then travel through the deep ocean, around Antarctica, perhaps north into the Indian or Pacific Oceans, and eventually return to the surface a millennium or so later.

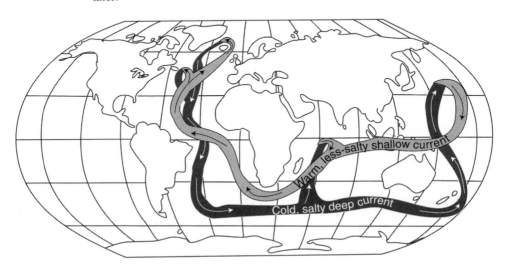

needs land to "lean on" so that the Coriolis effect doesn't turn the flow. The narrow north Atlantic has plenty of continents for the water to lean on, and warm surface waters can flow far to the north. But warm surface waters can't cross the Drake or the other, broader regions of the Southern Ocean. The flow south to the Antarctic is only efficient more than a mile down, deeper than the floor of the Drake Passage, where the water can "lean on" a rise in the sea floor and so move south.

The mile-deep ocean waters that flow to the Antarctic are already rather cold and salty. These cold, salty waters rise to the surface in the Antarctic, become a little colder by giving up a little heat to the atmosphere, and become a little saltier because freezing of the surface ocean makes low-salt ice and so leaves some extra salt in the water. This colder, saltier water then sinks back into the deep ocean. So while deep waters of the world's ocean are formed in both northern and southern regions, the effect on the atmosphere, and so on the world's climate, is much bigger in the north, where warm water cools, than in the south, where already-cool water cools a bit more.

Thus, Earth's climate works to move heat from the equator to the poles. Both the atmosphere and the oceans participate, and move similar amounts of heat. But because the ocean circulation depends on salt as well as temperature, the ocean heat transport is able to change drastically, as we will see next.

14

Conveyor belts in grocery stores usually work quite well, but occasionally one will fail. The great oceanic conveyor can also fail, tipping the world into new and unexpected climates.

Suppose you wanted to stop a grocery conveyor. You could locate some of the store's special, limited-time-only, free-with-$40-purchase flatware, grab a fork, and try to jam it into the mechanism. A good plan is to stick the fork into the gap where the conveyor belt goes down. The conveyor will try to pull the fork down but get it stuck, which may cause the conveyor to bind up and stop.

In the same way, the easiest place to influence the global ocean conveyor is the downgoing site, the north Atlantic. If the surface of the north Atlantic were a bit less salty, its water could not get cold enough to sink—it would chill and freeze without ever becoming as dense as the deeper waters. Then, warm water wouldn't flow north to replace the sinking water, European winters would switch from warm and wet to cold and dry, and polar bears might replace roses in the north of Great Britain. Just as freezing the U.S.-Canadian Great Lakes will chill Buffalo, freezing the north Atlantic would make Iceland much icier.

The north Atlantic is actually in a delicate balance at all times. The tropical ocean sends hot, salty water to the north Atlantic, but the tropics also send water vapor north in the atmosphere to rain or snow on the land and ocean. After the salty surface water leaves the tropics heading north, it gains more fresh water from rivers and rain than it loses to evaporation. The surface water is in a race—if cooling is faster than freshening, sinking happens in the far north; if freshening wins, the water gets stuck on the surface.

Suppose that a little more fresh water were delivered to the north Atlantic, because of more rainfall, or ice melt, or river runoff. Many researchers have built computer models of the ocean-atmosphere system. These models typically agree in showing that extra fresh water could "jam" the conveyor, greatly slowing or stopping it for a while before it restarted, and causing large, abrupt, widespread changes in the atmosphere that are very similar to those recorded in the ice cores, tree rings, and other sediments of the world. Other researchers have used subtle clues in ocean sediments to track the history of the conveyor strength. Many of these records indicate that the conveyor has weakened or stopped at times when the ice cores record sudden coolings in Greenland and elsewhere, and that the conveyor restarted when Greenland warmed. We will look at models of the conveyor next, and then at conveyor history.

Modeling the Conveyor

To model the behavior of the ocean, an ice sheet, the atmosphere, or almost any other system, you must first learn how that system works. You also must figure out what might cause the system to change, and what state the system is in at some time. These pieces constitute a model, which you can use to predict how the system will change. A wise person always tests a model to see if it works. Testing requires either predicting the future and seeing if the predictions come true, or starting the model sometime in the past and seeing whether (without cheating by looking at the answer) the model can "predict" the things that you know have happened. Hence, modelers like to know the history of whatever they are studying.

There are many ways to make models. The Biosphere II, for example, was a well-publicized attempt to build and study a model Earth in a big glass bubble in the Arizona desert. But the Biosphere was wildly expensive, and didn't do an especially good job of mimicking Earth. (It is a wonderful laboratory for other purposes, but it isn't the whole Earth.) The most affordable way we know of to model large pieces of Earth is to write equations for how we believe the Earth system behaves, and solve those equations on computers. If it bothers you to place any trust in computer models, remember that computer models have been used extensively in the design and testing of most of our cars, airplanes, buildings, bridges, and bombs for quite some time, with great success. Computer models can fail spectacularly, but when used wisely, they are valuable tools.

Many workers have built such models of the oceans and the atmosphere. These models probably are not as advanced yet as are models of bridges, but researchers are doing good things with the climate models. These models usually show that too much extra fresh water in the north Atlantic can have a significant effect. When only a little fresh water is supplied to the north Atlantic from continents or the atmosphere, the north Atlantic water is salty and sinks rapidly, allowing much more hot ocean water to flow into the north. When more fresh water is supplied to the north Atlantic, the water there sinks more slowly. But if the sinking slows too much, and the surface waters of the north Atlantic become too fresh, the sinking stops altogether.

This is a sort of "catastrophe." Once the sinking has stopped, the atmosphere continues to deliver fresh water to the north Atlantic, but this fresh water is not removed efficiently; instead, it puddles on the surface and may freeze in the winter. The sinking is then very hard to get started again; lowering the fresh-water supply a little bit still leaves fresh water pooled on the surface and the circulation stopped. You may have to make the nearby ocean really warm and salty before the sinking can start, bringing heat near the frozen puddle to melt it and allow the hot, salty water back to the far north, where it can cool rapidly during the next winter. Once the sinking stops, the ocean must take a long, cold route to return to its original state of vigorous sinking and warm land around the north Atlantic.

Any situation in which you can go back to where you started, but you must take a different route, is called a *hysteresis loop*. We see such loops every day. For example, consider driving into town on a two-way street toward a town square with a couple of one-way streets. Anywhere on the two-way street approaching or along the square, you could pull into a friendly driveway, turn around, and go back the way you came. But once you reach the far side of the square and turn onto a one-way street, you must go all the way around the square to get back to where you started, as illustrated in Figure 14.1. If you drive in on the sunny side of the square, and must go through the snowdrifts on the shady side before getting back to the warmth again, you have an idea of what happens to northern Europe when the north Atlantic water becomes too fresh to sink.

Models indicate that when the sinking stops in the north Atlantic, nearby regions cool a great deal and most of the hemisphere cools at least a little. The temperature difference between the equator and the far north increases, driving stronger winds that can carry more dust. The stronger winds also move surface waters of the ocean more rapidly and increase upwelling of cold waters in many places, including some tropical areas, cooling the surface ocean in those places and reducing evaporation of greenhouse gas water vapor from them. The cold north Atlantic reduces the rising of summer-heated air over Africa and into Asia that drives their great summer monsoons, so Africa and Asia become drier. A colder north and drier Africa and Asia reduce global wetlands, which would cause methane production to drop.

These and other changes in the models match the observed changes very well (although the observed changes have been somewhat bigger and more widespread than the models predict, a point to which we will return). In addition, ocean sediments show that the conveyor has slowed or stopped just at those times when the world switched into a

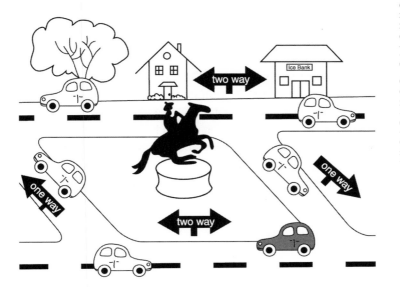

FIGURE 14.1

Hysteresis around the town square. Suppose you are the shaded car, driving into town. Anywhere along the two-way street, you could turn in a driveway or make a U-turn and return the way you came. But if you turn right onto the one-way street, you can't return the way you came, but must drive all the way around the square to get back to where you started. Such a situation is called a hysteresis loop. Today, if a little fresh water is added to the north Atlantic, the circulation slows a little; if a little fresh water is subtracted, the circulation speeds up on a two-way street. But if too much fresh water is added, the circulation stops, more fresh water from rivers and rain and ice melt piles up, and removing a little fresh water won't restart the circulation. The north Atlantic is then on the one-way street, and it is a long path to get back to the original vigorous circulation.

cold, dry and windy mode. Reading the ocean-sediment records to learn when the conveyor shut down is a neat trick, as we will see next.

On the Trail of the Conveyor

Even today, in the vastness of the ocean, it is difficult to directly measure the great currents of the conveyor. And tens of thousands of years ago, there certainly were no people with fancy current meters tracing the motion of the deep ocean. Instead, we have to look to clever interpretations of the ocean sediments to learn which way the currents were flowing then.

Water from different parts of the world's oceans has distinctive chemical and isotopic characteristics, which are picked up in certain key places, such as the north Atlantic, and then carried long distances before the chemistry and isotopes change much. The shells of plants and animals that grow in the ocean pick up signs of the water's characteristics. By studying the shells from different ages, we can learn how the currents changed. And change they did.

The open ocean is almost a biological desert. Shallow waters have plenty of sunshine, water, and carbon dioxide with which to grow algae rapidly, yet rapid growth is rare in most places because the living things also need a little phosphate, nitrate, iron, and other "fertilizers" that are rather scarce in the open ocean. Where such nutrients do occur, algae bloom. But animals quickly eat the plants, and package some of the plant remains into fecal pellets that sink into the deep ocean, where sunlight is not available to fuel growing algae. We saw earlier that this removes carbon dioxide from the surface ocean and ultimately from the atmosphere, but it also removes the nutrients in the plant remains. As the nutrients are removed from the lighted zone, the bloom ends unless more nutrients are supplied.

Some of the nutrients are buried in ocean sediments, and eventually may be taken deep into the earth by continental drift processes, melted, and returned to the surface by volcanoes. Many of the nutrients, and much of the carbon packaged with the nutrients, dissolve in the deep ocean as the pellets drift through it, or as burrowing creatures stir up the sediment while looking for something to eat. Over time, deep-ocean water accumulates materials dissolved from sinking pellets, whereas surface waters are rapidly stripped of such materials to make the pellets.

Cadmium is one such material that accumulates in deep water over time; this is useful because a little bit of cadmium sneaks into calcium carbonate shells in place of calcium. The isotopes of carbon are similarly useful. Algae find it "easier" to use the light isotope of carbon, leaving heavier carbon in the surface water. Pellets thus transport light carbon, which is released into the deep ocean, where it is used in calcium carbonate shells. If the shells of bottom-dwelling animals are low in light carbon and cadmium, this indicates that the shells grew in water that recently had been at the surface. If the shells of bottom-dwellers are rich in light carbon and cadmium, this means that the animals were living in water that had been in the deep ocean for a long time—a millennium or more.

Oddly enough, the waters that rise to the surface around the Antarctic and then sink again don't spend a very long time at the surface, often remain beneath sea ice or ice flowing off Antarctica, and so don't fully lose the signature of the deep ocean. But waters that sink in the north Atlantic gain the chemical and isotopic signatures of their long trip north along the surface.

Today, water sinks in the north Atlantic in two general regions: to the north and east of Greenland in the Greenland, Iceland, and Norwegian seas, and to the south and west of Greenland in the Labrador Sea. Work is ongoing to learn the details of how this pattern of far-north and not-so-far-north sinking has changed in the past, but a likely story is emerging. The north Atlantic circulation has three modes: the hot mode, with far-north and near-north sinking; the cool mode, with only near-north sinking; and the cold mode, with no northern sinking.

Looking back in time, north Atlantic waters from warm times were much like those of today. During cold times, however, the water that filled the deep north Atlantic had been in the deep ocean accumulating nutrients and light carbon for much longer than under modern conditions. During the coldest millennia of the ice age, and during each of the cold times of the Dansgaard-Oeschger oscillations, sinking of waters seems to have slowed greatly or stopped in the far north. Much sinking continued in the near north during most cold times but wasn't quite as fast and didn't reach quite as deeply as during warm times, allowing waters from the Antarctic to spread north along the very bottom of the Atlantic. During the coldest times of the Heinrich events, sinking stopped almost entirely in the far and the near north Atlantic, allowing older, Antarctic-origin waters to fill much of the Atlantic.

The models show that as warm waters penetrate closer to the poles, they have a greater effect on the atmosphere. Hence, the shutdown of the far-northern sinking during Dansgaard-Oeschger coolings affected the atmosphere greatly, and the additional shutdown of the near-northern sinking during Heinrich events caused only a little more cooling. But the additional Heinrich-event shutdown had a greater effect on the ocean, because the warm winters of modern Europe are "stolen" from the south Atlantic.

Stealing from the South

Much of the surface water of the Atlantic Ocean is supplied from the Pacific or Indian oceans, passing into the great Southern Ocean below Africa and South America and then into the Atlantic. This water flows north on the surface of the Atlantic, moving into and through the tropics to the north Atlantic before sinking, flowing south, and eventually upwelling in the Pacific, Indian, or Southern oceans and returning. The Atlantic surface waters are warmed in the southern as well as the northern tropics, but release the heat in the north. If the northward flow of Atlantic surface water were to stop, the southern heat would remain in the south, warming the south Atlantic as the north Atlantic cooled.

During the cold phases of most Dansgaard-Oeschger os- cillations, when sinking largely stopped in the far-north but continued in the near-north Atlantic, some of the sun's heat was still taken north across the equator in the ocean currents. This heat was released in not-so-cold places that receive a lot of sunshine, however, so this ocean heat didn't greatly affect the atmosphere, allowing the far north to cool greatly. But during Heinrich events, the fresh water from the melting ice- bergs seems to have stopped or greatly slowed all of the north Atlantic sinking. This double shutdown caused larger and more widespread cooling, drying, and wind-speed increase than a single shutdown, but the changes were not doubled, and their patterns in the northern hemisphere were similar to those for a single shutdown (see Figure 14.2).

The heat left in the south Atlantic during a double shutdown would be expected to cause a warming there. In addition, the double shutdown may have increased the southern sinking of waters, giving a little more southern warming. Mixing processes in the oceans cause the deep waters to slowly become warmer and less salty until they rise toward the sur-

FIGURE 14.2

A diagram showing the three-mode view of the ocean. During HOT times, warm water gives its heat to the atmosphere and then sinks in the far north Atlantic and the near north Atlantic. During COOL times—all of the colder parts of the Dansgaard-Oeschger and Heinrich-Bond oscillations, and the few millennia when orbital effects caused the heart of the ice age—the far-northern sinking seems to have largely or completely stopped, but the flow of warm waters across the equator to the near-northern sinking continued. During the COLD times of Heinrich events, the near-northern sinking also shut down, leaving hot waters in the southern hemisphere to warm the south Atlantic.

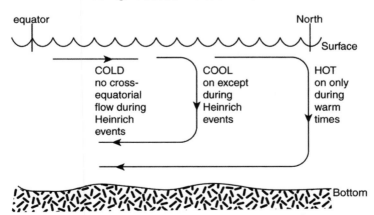

Atlantic Ocean currents
during hot, cool, and cold times

face and are replaced by more deep water, so deep water must be made somewhere. A north Atlantic shutdown would cause, within a few decades or centuries, an increase in deep-water formation somewhere else, probably in the Atlantic sector of the Southern Ocean, where much deep water already forms. And while Antarctic deep waters are made from rather cold waters from deeper than the floor of Drake passage, these waters cool a little more in the Antarctic, giving a little heat to the atmosphere in the process.

The single shutdowns of the Dansgaard-Oeschger coolings between Heinrich events were enough to affect the atmosphere of much of the northern hemisphere, and even across the equator a bit, but did not do too much at high southern latitudes. The double shutdown of a Heinrich event sent a larger cooling signal through the atmosphere from the north toward the south, but also sent a warming signal into the south Atlantic. These mixed messages apparently created the complex mosaic we observe, in which some southern regions centered on and downwind of the south Atlantic warmed during Heinrich events while other southern regions cooled. Apparently, the south Atlantic was "listening to" the changes in the ocean currents, while other regions were "listening to" the atmospheric signal transmitted from the north Atlantic into and through the tropics.

Cataloging the Conveyor

So the picture is becoming much clearer. The salty Atlantic surface waters of the ocean conveyor belt pick up heat from the south Atlantic and deliver it to the north Atlantic, either far to the north or not quite so far to the north. There, the heat is released to the atmosphere to warm the winters, allowing the water to cool and sink to the deep ocean. If too much fresh water is supplied to the north Atlantic, the water there will not sink and the winters of Europe become cold.

Shutting off the far-northern sinking to go from the warm mode to the cool mode is fairly easy. Far-northern sinking has been shut down for a few centuries during most millennia, extending back perhaps a million years or more. Shutting off this far-northern sinking has had large effects on the atmosphere. Colder air typically carries less water, so each shutdown dried many parts of the northern hemisphere. A shutdown also weakens the monsoon circulation that brings moisture to large parts of Africa, and may weaken the Asian monsoon as well—north Atlantic coolings caused the Sahara Desert to expand, shrinking the tropical wetlands that produce "swamp gas" methane and so lowering the methane concentration of the atmosphere.

The cooling and drying during shutdowns caused large changes in the types of plants and animals living in widespread parts of the northern hemisphere. Temperature differences help drive air motion, so strong northern cooling caused stronger winds, dustier skies, and other changes. Stronger winds also help mix the ocean better, cooling the surface waters of tropical oceans, reducing the greenhouse gas water vapor in the tropics, and so causing some additional cooling that reached into the southern hemisphere. However, most of the changes from a far-northern shutdown were restricted to the northern half of the globe, and especially to regions in and around the north Atlantic basin.

The far-northern sinking apparently stayed shut off for many millennia when Earth's orbital wiggles had caused the coldest conditions in the north, with ice sheets reaching their greatest extent. While the far-northern sinking stayed "off," the climate was somewhat stable, although in an ice-age pattern quite alien and unfriendly to us. Similarly, the far-northern sinking has remained on for the last few thousand years, when orbits have caused the northern climate to remain warm and the ice sheets to remain small. For the rest of the record, the far-northern sinking has flipped on and off, causing large jumps between a warmer-wetter-calmer climate and a colder-drier-windier one in the northern hemisphere, with each cold-warm Dansgaard-Oeschger cycle lasting about 1,500 years.

During this millennial on-and-off flipping, successive cold jumps have gotten colder and colder as ice sheets grew in

Canada and Europe until a Heinrich-event surge of the Canadian ice from Hudson Bay shut down all the north Atlantic sinking and switched the north Atlantic from the cool mode to the cold mode. The additional shutdown did not have too much extra effect on the northern atmosphere, which is made incredibly cold just by a single shutdown, but the double shutdown cooled the atmosphere in broader regions. The extra shutdown stopped the ocean-current theft of heat from the south Atlantic, and probably caused additional sinking of waters around Antarctica. While much of the world was experiencing conditions even colder than those for the single shutdown, regions centered on and downwind of the south Atlantic warmed in a sort of climate seesaw.

A small but important additional complication merits mention here. The millennial flipping seems to have continued during the warmest and during the coldest times, but with tiny effects, rather than the usual huge ones. Perhaps the climate is like a theater with orchestra, mezzanine, and balcony levels: You can move up or down a bit between rows in the orchestra level, but then there is a big gap to the mezzanine level, different rows there, and then another big gap to the different rows of the balcony. Sometimes the climate changes a little between rows; sometimes it takes big jumps across the gaps. Today we live in the hot balcony of the north Atlantic theater, moving up and down among the balcony rows. During most of the last 100,000 years, and probably the last million years or longer, Dansgaard and Oeschger have chased the climate between balcony and the cool mezzanine, although the climate sat in the mezzanine during the coldest phase of the orbitally induced changes. Heinrich events kicked the climate from the mezzanine all the way down to the orchestra level, giving the coldest north Atlantic but warming the south Atlantic.

15

Why must the climate keep jumping among rows and levels, rather than settling down to watch the concert? We have a good story for the Heinrich events, with changes in ice sheets causing sudden delivery of fresh water to the north Atlantic and stopping the conveyor. For other events, however, we really don't know. Volcanoes, the sun, floods, and other causes have been suggested. Some of these can now be eliminated—it probably wasn't the volcanoes—and others may be part of the answer—the floods matter.

A large volcanic eruption puts enough sulfuric acid into the stratosphere to block a noticeable amount of sunlight for a year or two and cool the globe by a degree or so during that time. But volcanoes are scattered across Earth and don't have very good ways to get organized and erupt at the same time. Volcanic effects on the climate thus usually are more random than periodic. Ice-core records don't show a good, consistent correlation between abrupt climate changes and the fallout from volcanic eruptions. We thus believe that volcanoes are not very important in our story.

The sun's energy output varies a little bit—about 1/10 of 1 percent—over the eleven-year sunspot cycle, and many climate records show very small changes in response to the altered sunshine. Somewhat longer sunspot cycles of a century or so have included the "Maunder Minimum," a few decades around the year 1700 when sunspots were very scarce, and some estimates are that the sun's energy output then may have been reduced 2/10 of 1 percent compared to typical modern output. (Surprisingly, the dark sunspots are associated with a hotter sun, because bright bands around the sunspots give out more additional energy than the dark sunspots block.) Based on somewhat sketchy data, the Maunder Minimum appears to have been the center of a couple of centuries of generally low solar output. Interestingly, this time of reduced solar activity corresponds more or less to the "Little Ice Age," the few centuries of cold that preceded the warming of the twentieth century. Might the sun have varied over about 1,500 years to cause Dansgaard-Oeschger changes?

More data would help us answer that question, but the indications now point away from the sun. The evidence comes from our ice cores. When the sun is more active, the solar wind does a better job of protecting Earth from cosmic rays. Cosmic rays bash molecules in the atmosphere and make new types of atoms, such as beryllium-10; those atoms then fall on Earth and its ice sheets. A brighter, more active sun lowers beryllium-10 production, while a dimmer sun allows cosmic rays to make more beryllium-10.

The concentration of beryllium-10 in ice shows a very nice eleven-year sunspot cycle and century-length changes such as the Maunder Minimum, and the ice cores also record the small changes that the varying sun has caused in temperature and other aspects of the climate. The concentration of beryllium-10 in ice changed significantly during the big climate jumps, but this is largely explainable by the effect of changes in how much snow fell on the dilution of the beryllium-10 in that snow. And during the current warm period, the ice cores do not show much change in beryllium-10 production over thousands of years. If the sun was involved in the abrupt climate events, then solar changes too subtle to show up easily in the beryllium-10 records must have been amplified hugely in the Earth system.

Beryllium-10 production also is affected by the strength of Earth's magnetic field. About 40,000 years ago, a weakening of the magnetic field seems to have caused an increase in beryllium-10 concentration in ice cores and other sediments. This event does not correlate with any climate changes, however, so the magnetic field doesn't seem to be important in the climate story either. We will continue to look for evidence, but the answer to the millennial variations in our climate likely must be sought here on Earth. Fortunately, the history of giant floods gives us a better idea.

Fickle Floods and Oscillating Oceans

The ice that advanced from Canada into the United States during the most recent ice age blocked rivers that flowed north and east. The entire drainage of the U.S.-Canadian Great Lakes was plugged up, and the water formed ever-larger lakes until they overflowed down the Mississippi River. When the ice began to melt, eastern outlets were opened down the Susquehanna, and then the Hudson, and eventually the St. Lawrence rivers. Small advances and retreats by the ice edge, in response to climate changes or processes internal to the ice sheets, opened and closed some of these outlets more than once.

No reasonable person would want to dam a river with ice. Once a significant amount of water starts to run over, under, or through an ice dam, the turbulence in the flow makes "frictional" heat that melts more of the ice, allowing more water to flow through. Ice dams fail catastrophically. The largest floods known on Earth have been caused by failures of ice dams.

Each time that retreating ice opened an outlet through which ponded waters could drain to the east, a huge flood would have occurred as the lake dropped to the new level, followed by a sustained fresh-water supply fed by the water flowing into the lake. Dumping a large amount of water into the north Atlantic in a hurry is a great way to stop the conveyor circulation and chill the north Atlantic, and continued fresh-water supply would help keep the conveyor turned off. A huge flood down the St. Lawrence River occurred just be-

fore the Younger Dryas cooling, and these two events probably are related. Some earlier coolings also occurred just after
meltwater was dumped eastward into the Atlantic.

Further evidence strengthening this story is that the most
recent large, abrupt cooling has been linked to a meltwater
flood. This event, which occurred about 8200 years ago,
caused climate changes that were about half as "big" as those
of the Younger Dryas, for about one-tenth as long. This occurred just after the last ice melted out of Hudson Bay, allowing the large lakes that had puddled around the ice to drain
through the bay to the north Atlantic.

The clear evidence linking some abrupt changes to sudden meltwater deluges is tantalizing—maybe all the abrupt
changes were caused this way. However, there have been
many abrupt changes, probably hundreds over the last million years, with similar spacing in time. Most scientists doubt
that meltwater outbursts would have been so frequent and so
regular.

The millennial scale of the Dansgaard-Oeschger oscillations then focuses our attention on the ocean, rather than on
glaciers and ice sheets or the atmosphere. The big ice sheet in
Hudson Bay took many thousands of years to change, as
shown by the spacing of the Heinrich events, and the rest of
the Canadian ice sheet took even longer to change, so this ice
probably wasn't changing much every thousand years as well.
Smaller ice sheets could have changed more rapidly, but their
small size would have made it difficult for them to release
enough meltwater to the north Atlantic to have much effect
on its circulation. And the small ice sheets suffer from the
same problem as the volcanoes—they have trouble getting
organized to work together. The atmosphere changes over
weeks or months, and cannot remember what it was doing
long enough to cause thousand-year cycles.

In ocean circulation, on the other hand, water takes about
one thousand years to make a loop of the conveyor belt, so it
seems reasonable to suspect ocean changes. Wally Broecker
and his coworkers proposed one way that this might work.
When the ocean conveyor provides a lot of heat to the north

Atlantic, ice on nearby land will tend to melt, dumping fresh water into the north Atlantic. Eventually, the fresh water might stop the conveyor circulation, or at least the far-northern sinking. The north Atlantic would cool, and ice might regrow on land. Ice growth on land requires that more water leaves the ocean than returns to it, so the north Atlantic Ocean would become saltier, perhaps starting the sinking again. Models indicate that such a cycle could occur over about a millennium. Changes in meltwater delivery from ice-dammed rivers could speed or slow the arrival of the next cooling, causing the irregular spacing of the events.

In all honesty, this story does not answer all the questions. For example, since the great ice sheets melted during the current warm period, the large ice sheet of Greenland and tiny glaciers and ice caps on some islands are all that remain in the north. Growth and melting of these ice masses might explain the small oscillations in climate that have persisted through the current warm period, but why hasn't the spacing between events changed as the climate changed?

Given all the remaining uncertainties, I expect that there will soon be more to our story—we really don't understand the Dansgaard-Oeschger oscillations yet. But the search is on.

Other Amplifiers and Switches?

The north Atlantic almost certainly was the center of action for many of the major climate changes of the past. Data and models agree that fresh water supplied to the north Atlantic can "flip the switch" of the climate, bringing wildly different conditions. But anyone who has wandered around a modern home knows that one light may be controlled by several switches, and other switches may operate additional lights, the garbage disposal, and the garage door opener. Might the north Atlantic respond to other switches? And might the climate in other regions of Earth have its own switches? Would this help to explain why changes of the past have been even greater than expected based on the effects of the north Atlantic?

These are important questions, and ones for which we lack good answers. Many tantalizing clues suggest that the answer to these questions is "Yes"—other switches and amplifiers exist in Earth's climate.

The north Atlantic has changed much since the ice age, but the weak climate oscillations of the current warm period have about the same millennial spacing in time as the larger climate oscillations of the ice age. This suggests that there is a millennial climate oscillation affecting the climate system from outside the north Atlantic. The big climate changes may occur when the "accidents" of outburst floods and ice-sheet surges in the north Atlantic add to the millennial cycle from elsewhere. But where is the cycle coming from? The deep ocean? The tropics? The south?

The large, energy-rich tropics are the heat engine that drives Earth's climate. Changes in the tropics are easily transmitted to much of the world, as shown by the widespread mayhem created by El Niño. During El Niños, anomalously hot water along the equator in the Pacific "steams up" the atmosphere, giving unusually warm and wet conditions in many regions but drying other areas as weather patterns shift. Flood and drought, landslide and fire result. A switch in El Niño frequency would have profound effects on climate.

Unfortunately, long records of tropical conditions with high enough time resolution to detect El Niños have been hard to find. Wonderful ice cores from high mountain glaciers in the Andes reveal changes in El Niño over a few millennia, but the thin ice and high snowfall rates have caused layer thinning to be so extreme that longer records reaching back to the ice age lack sufficiently high time resolution to see El Niños. These and other records provide tantalizing hints about changes in El Niño, but certainty is elusive.

Much of the deep water in the ocean sinks in the vast and poorly known southern ocean around Antarctica. The complex interplay of ocean, sea ice, Antarctic ice sheet, and atmosphere there may hold many surprises, and many of us in the climate-change community are now focusing our efforts on the Antarctic.

Until better records are available, we are left to speculate whether changes in El Niño, in other parts of the tropics, in the Antarctic, or even in the sun or something else served to cause or amplify the abrupt climate changes in the past. It may not be too far-fetched to think of the climate researchers confronting abrupt change as a Stone Age tribe clustered around the north Atlantic light switch, the first one we've ever seen, and just now figuring it out. But some of our number are beginning to set off to see what runs the garage door. The north Atlantic switch is undoubtedly important, but most climate researchers expect the story of abrupt climate changes to become more complicated before it is finished.

V

COMING CRAZINESS?

What might

happen to

Earth's climate

in the future—

and what we

might do

about it

16

Ice cores and other sediments show that large, rapid, and widespread climate changes have been common on Earth for most of the time for which we have good records, but have been absent during the critical few millennia during which agriculture and industry arose. At least some of those large changes appear to have been triggered by increased fresh-water delivery to the north Atlantic. Climate jumps have been especially common when changes were occurring in important parts of the climate system, including summer sunshine in the north, carbon dioxide in the atmosphere, and ice-sheet size.

The critical questions for us are: Will nature, or humans, return the climate to the "normal" condition of wild jumps rather than the "anomalous" stability that we now enjoy? And, if such a return seems likely, is there anything we can do about it?

Firm answers are not available, unfortunately, and I doubt that answers will be found in the immediate future. But there is a significant possibility that greenhouse warming could trigger enough extra rain, snow, and ice-sheet melting to partially or completely shut down the north Atlantic conveyor circulation. Greenhouse climate changes thus could be larger and stranger than most people expect, including wintertime freezing around the north Atlantic.

In this chapter, we will consider why the climate-change community is so much more confident of global warming than is the popular press. We will also take a short detour into economics, to see why abrupt climate changes are the most important ones, and why businesspeople and environmentalists may yet end up on the same side of the argument.

Where the Gas Goes

Earth has an amazingly efficient recycling system. Plants capture and store sunlight by using its energy to make more of the carbon dioxide/water combinations that we know as plants. But animals, fungi, and bacteria make more of themselves by rearranging the plants they eat, or they slowly burn the plants they eat to gain the energy of the stored sunlight. Almost everything that dies is recycled within a few years of its death.

The recycling is not perfect, however, and a few dead things escape recycling for a while, "leaking" out of the usual cycles. Where plant decay is slow (as in cold tundra regions) or where the supply of dead plants is very large (as beneath regions of ocean upwelling, where nutrients from the deep ocean fertilize blooms of plankton), dead plants may pile up faster than they are burned. Continued burial of piled-up dead plants moves them deeper inside Earth, where geothermal energy "cooks" them. The result is a fossil fuel: oil mostly from algae, coal mostly from woody plants, and natural gas—mainly methane—from either one.

This "leaking" has been going on for hundreds of millions of years, turning carbon from volcanoes into fossil fuels in rocks. Humans have discovered that it is easy to collect and burn this fossil fuel, turning it into energy we want and carbon dioxide that we release into the atmosphere. We are in the middle of a few centuries of easy living fueled by a few hundred million years' worth of stored solar energy that evaded earlier recycling.

Some of the carbon dioxide we release into the atmosphere dissolves in the ocean, some goes to grow trees, but much stays in the atmosphere for a while. We can hope mightily that most of the carbon dioxide we release will go somewhere that it won't bother us, but this is not likely to happen. We probably are burning trees as fast as, or faster than, new ones grow. The carbon dioxide going into the ocean is slowly changing the water chemistry, making it more difficult for more carbon dioxide to enter the ocean. (It is difficult to put more cats in a kennel, to heat an already-hot pan, or to add more carbon dioxide to a carbon dioxide-rich ocean, because of the tendency for most things to spread out from where they are concentrated.)

Carbon dioxide and water combine to form a weak acid that reacts with and dissolves rocks. Seashells and coral-reef skeletons are special rocks that living organisms build around themselves. Making the ocean more acidic will make shell growth more difficult, and will tend to dissolve the shells of dead creatures. But dissolving shells temporarily neutralize some carbon dioxide, which is part of the reason why some of our carbon dioxide goes into the ocean rather than building up in the atmosphere. If we put too much carbon dioxide into the air, the oceans will begin to run out of shells to dissolve and so will have greater difficulty absorbing carbon dioxide, and most of the carbon dioxide we produce will stay in the air for centuries, millennia, or longer.

During ice ages, extra carbon dioxide was taken up by the ocean, in part because the colder water then could hold more gases, and in part because extra dust from the stronger ice-age winds fertilized plants that used more carbon dioxide. In the future, warming will work against us, releasing carbon dioxide from the ocean to the atmosphere. Some people have suggested that we humans could fertilize the oceans, duplicating the natural feat of the ice-age winds. This is indeed possible, although studies indicate that the amount of carbon dioxide likely to be taken up in a fertilized ocean is small compared to the amount we are planning to release by burning fossil fuels. Also, if too much carbon dioxide were turned into algae that sank, the decay of the algae could deplete the deep ocean of oxygen. This could cause extinctions there, and might allow release of other greenhouse gases, such as methane or nitrous oxides.

In short, most of the carbon dioxide we put into the atmosphere will stay there for a while, acting as a greenhouse gas. And the more carbon dioxide we release, the more will stay in the atmosphere. The few centuries of fossil-fuel burning thus will produce a few millennia or tens of millennia of elevated atmospheric carbon dioxide concentrations.

Eventually, the carbon dioxide we release will combine with volcanic or other rocks in the slow process of weathering. The weathering products including the carbon dioxide will be washed to the ocean, and (with difficulty) turned into corals or clamshells. A little bit of the carbon dioxide used by algae will escape being recycled and begin to form new fossil fuels. The atmosphere and the ocean will "forget" what we have done over tens or hundreds of thousands of years. A few hundred million years from now, when new fossil fuels have built up in new rocks, even the geology will forget us.

For the coming millennia, while the atmosphere is enriched in carbon dioxide, however, the planet will be warmer than it would have been with less carbon dioxide. This is the human enhancement of the natural greenhouse effect, which the press usually shortens to "the greenhouse effect." Almost everyone—environmentalists and industrialists, right and left, first- and third-worlders—agrees that increasing the carbon dioxide in the atmosphere will warm the planet at least a little if all other things are held constant. Carbon dioxide in the atmosphere has little effect on incoming sunlight, but blocks some of the longer-wavelength radiation that Earth sends back to space. If carbon dioxide increases, incoming energy will exceed outgoing energy, and the difference will cause warming to a level at which Earth is able to force enough energy past the carbon dioxide to balance the incoming sunlight.

Disagreements start very soon after this, however. The direct radiative effect of the human-caused increase in carbon dioxide is not likely to be too large—something vaguely in the neighborhood of one degree of warming over the next century. But, remember how chock-full of feedbacks the Earth system is. Most climate researchers expect the feedbacks to amplify the changes, giving several degrees of warming over the next century.

For example, warmer air can hold more water vapor, which is a more potent greenhouse gas than carbon dioxide. Warming probably will melt some of our highly reflective snow and ice, leading to additional absorption of sunlight and, thus, to additional warming. Today, the short plants of the tundra of the far north can be buried by snow that reflects sunlight, whereas the dark trees of the taiga just to the south stick through the snow and absorb what sunlight is available; shrinkage of the tundra with warming thus may cause further warming.

These feedbacks are less certain than are the direct radiative effects of carbon dioxide. Some people emphasize the difficulties in predicting the feedbacks. For example, there are hot places without much water vapor (deserts, for example), so maybe warming will not increase water vapor greatly. Also, more water vapor will make more clouds, and while some clouds (the high, thin ones) warm Earth by blocking more outgoing than incoming energy, others (the lower, thicker ones) cool the planet by blocking more incoming than outgoing energy. If an increase in carbon dioxide causes an increase in low, thick clouds, the total temperature change may not be very large. Snow melts in many places when warming occurs, but some places are cold enough that warming will not melt their snow, and may even bring more snow because warmer air often supplies more moisture.

The modern scientific consensus is that positive feedbacks will amplify global warming. This is reflected in the Intergovernmental Panel on Climate Change (IPCC) reports, which are produced ultimately under the auspices of the United Nations and which represent the painfully forged results of discussions by thousands of scientists, government officials, interest groups, and private citizens.

Agreeing to Disagree

A word on "scientific consensus" may be useful here. *All* scientific ideas are subject to revision; we should never be absolutely sure that the truth has been reached. Old ideas should be tested continually, in an effort to tear them down and replace them with better ones. Ideas that survive this constant attack will be especially robust. Experience shows that if we then behave as if these surviving ideas are true, we will succeed—in curing diseases, finding clean water, building things that stand up when we want them to or blow up when we want them to, and so on. But, on the other hand, the ideas may be true, they may be reasonable approximations of the truth, or we may just be lucky.

Because there is honor in tearing down old ideas to replace them with something better, science wants and needs contrarians who hammer away at the old ideas. So "scientific consensus" is not the same as 100 percent agreement, and never should be.

Add to this the fact that huge money may rest on the global-warming debates. If the United States or the world decided to change the tax codes to reduce fossil fuel use, solar-energy startup companies would become more valuable, while oil wells would become less valuable. If we decided not to worry about global warming, many researchers and many lobbyists might suffer. When real money (which may add up to tens or hundreds of billions of dollars) reinforces the scientific need for contrary ideas, you can be absolutely sure that there will be loud voices on all sides of an issue. In the interest of fairness, the political process and the press tend to further confuse casual observers by giving a contrarian voice the same stature as a near-consensus voice.

Some searching with an open mind is required to sort through the loud voices and identify the leading ideas. I have tried to do so (and you can judge whether I have succeeded), and I believe that the weight of scientific evidence indicates that significant human-induced future warming is the most likely outcome.

Some loud voices focus on the possibility of natural changes opposing this human-caused warming. This view is absolutely right; natural changes could offset some or all of the human-caused warming. However, there is an approximately equal chance that natural changes will go the other way, greatly increasing the changes that we cause. Because the human-caused changes are likely to be much larger than any natural changes that industrial or agricultural humans have experienced, it is not too likely that the natural changes will suddenly become large enough, and go in the right way with just the right timing and distribution across Earth, to offset what we humans do. (We do expect the natural trend to be a slow, 90,000-year cooling into the depths of a new ice age, but the globally averaged rate of cooling over that time would be something around 0.01 degree per century, and maybe three to four times bigger in the polar regions, where changes are largest. Human-induced changes are likely to be one hundred or more times faster, so the next natural ice age won't save us from ourselves. Some visionaries have even talked about saving the fossil fuels until we really need them to fight an ice age, but that would be so far in the future that it is difficult for human economies to deal with.)

Ice cores and other paleoclimatic records figure prominently in the global-warming debate, and indicate that rising carbon dioxide levels will cause significant warming. The oldest direct measurements of atmospheric composition are from ice-core samples from just over 0.4 million years ago, but indirect methods agree that the warmth of the saurian sauna of 100 million years ago was caused in part by high levels of carbon dioxide. Similarly, the best explanation of the faint-young-sun paradox is that Earth did not fall into a permanent deep-freeze billions of years ago because higher carbon dioxide concentrations then allowed warmth with less sunlight.

Over the last 0.4-million years, the Vostok ice-core record from Antarctica shows that temperatures and greenhouse gases have changed in similar ways. The carbon dioxide almost certainly is not the ultimate reason for the climate shifts, which were caused by orbital wiggles moving sunlight around on Earth. But the Antarctic cooled when Canada had short, cool summers even if the Antarctic was receiving extra sunshine. Credible explanations of this behavior all involve the greenhouse effects of carbon dioxide changes and associated positive feedbacks. Carbon dioxide and temperature records are certainly not identical—many things affect the climate—but the similarities of carbon dioxide and temperature records are unmistakable.

Note that the controls on carbon dioxide have been different on these different time scales. Over millions to billions of years, the important balance was between carbon dioxide consumption by chemical reactions with rocks and carbon dioxide production by volcanoes. Over the hundreds of thousands of years of ice-age cycles, the rate at which the biological pump moved carbon dioxide out of the surface oceans probably has been most important. Over the centuries that we will burn fossil fuels, these other processes will be slow compared to our actions, and we will be the most important control of carbon dioxide levels. But whatever the controls on these different time scales, warmth and elevated carbon dioxide levels have gone together for billions of years. It is highly likely that this relation will continue in the future.

A much tougher question is whether global warming will be a bad thing, and whether we should do something to slow or stop human influence on the climate. I offer a few of my thoughts on this in the final chapter of this book. I recognize the remaining great uncertainties, but generally support the international consensus embodied in the U.N.-sanctioned IPCC reports that the warming we humans cause will hurt some of us and help others, and that the hurt will probably outweigh the help. I believe that serious discussion is needed on human response to these changes, which may lead to human actions. This may require a slightly different view of the problem than used in traditional economic analyses, however.

Traditional economic analyses often suggest that we should do a little about global warming, but not a whole lot. This result comes from a peculiarity of the analyses, and from the slowness of projected changes.

If I ask you to give me a pile of money, you will insist that I eventually give you back a bigger pile of money. The extra represents several things: uncertainty (maybe I'll run away with your money); opportunity (your money would grow if you put it into a bank account or the stock market, so you expect your money to grow with me as well); and preference (you want the things you can buy with that money now, not far in the future). Because of these and other ideas (which, in some ways, are the same thing), an apple or a dollar is worth more to you today than it is in the future, and the further you look into the future, the less a dollar is worth to you. Economists call this idea *discounting*, and what discounting means to an economist is that we should deal with the uncertain future from climate change (and from all other causes) by building a big economy and then letting the economy handle whatever happens. With typical discount rates, things that happen more than a few decades in the future have little value, so one doesn't worry about them too much. Hence, it may be useful to do a little about slowing human effects on climate, but we shouldn't do a lot.

True, there are "environmental" types (including some economists) who view things differently. Ethics figure into the discussion—the changes we cause will last for millennia, so do we have the right to subject future peoples to those changes? Nontraditional discount rates also show up in the discussion. Today, for example, many people have money in low-yielding bank deposits. After you subtract inflation and taxes, these people are losing money every year. Some economists would say that the people with low-yielding bank deposits are just stupid, ignorant, or lazy. But perhaps these people are living fairly well now, have lived through the

Great Depression, and are really concerned that some day they could lose everything and wind up living in refrigerator cartons under freeway overpasses. An apple in the future may be worth more than an apple today to such people, because they have enough apples today and are willing to spend a little for security. Still, this is a minority view—most economists probably would argue that the slowness of climate change means that we should put most of our efforts into adaptation rather than prevention.

Other reasons can be advanced for nontraditional discount rates. Some economists may be using rather optimistic projections of how much the whole world's economy can grow in the future, and that may lead them to worry too little about future problems. Economics traditionally assumes that there is always a substitute—if something becomes scarce and expensive, someone will figure out a replacement for not too much more money. But Earth is finite, and the jump to space flight may be wildly expensive, so perhaps we should not assume that the economy can grow forever. Issues of fairness also come into play—climate change may hurt most people on Earth, but the fossil-fuel burning that contributes to climate change now is benefiting some people (those in the developed world who do most of the burning) more than others (those in the developing world). These are important questions and deserve careful scrutiny.

The paleoclimatic record has added an additional critical issue to the economic debate on global warming. Our ice-core records show that huge shifts have happened in the climate—not over centuries or even decades, but over years. Over just a few years, discounting does not affect the value of things very much. If we knew that a large climate change was coming, that the change would cost our economy a lot, and that altering human behavior could prevent the climate change (huge "ifs"!), then a traditional economist likely would agree that we should do a lot to avoid the change.

Now, to be perfectly honest, we don't know whether an abrupt change is coming that would freeze northern Europe or parch Africa. Even if we knew that a change was coming, we don't know whether we could do anything about it. And it is likely to be a while before we learn such things. So if you are looking for a ringing conclusion and a call to arms (or bicycles?) I apologize now, because there won't be one. But we do know that a large, abrupt climate change could happen, and that human activities may make large, abrupt changes more likely.

17

What are the odds that natural or human activities will trigger an abrupt climate change big enough, fast enough, and soon enough to matter in economic discussions? The simple answer, again, is that we do not know. The widespread realization that such an event is even possible is only a few years old. We continue to suspect that the known north Atlantic "light switch" is only one of several such switches in the climate system, but we aren't even sure about this. Much knowledge is needed before we can begin to predict the known light switch, and it remains possible, though unproven, that "chaos" in the system will render such predictions difficult or impossible. The study of abrupt climate changes really is in its infancy. My rallying cry is: "Send your brightest students to help, and cheer them on!"

Meanwhile, however, let's have a look at any clues to recurrence in the climate record, and at what the existing models say about the possibility that humans will help flip a switch. The indications here are not favorable—both nature and humans are capable of flipping a switch.

Staggering Naturally

The climate records show that over the last 100,000 years, the climate was most "boring" during the coldest 10,000 years and during the warmest 10,000 years, with jumping climates during the coolings and the warmings. (And remember that even during the most boring times, there were still changes that have helped nurture and then topple empires.) A possible lesson would be that we should try to keep the climate really warm or really cold, away from the middle ground where jumps occur. Given that the orbital trend should be a long, slow, bumpy 90,000-year slide to the coldest part of the next ice age, one might suppose that warming would help prevent abrupt changes.

But the warmest and coldest times were also the times when the climate wasn't being forced to change much. In the roller coaster of orbitally changing sunshine distribution, the boredom of the nearly flat hilltops and nearly flat valleys were separated by climbs and drops. Greenhouse gases followed the sunshine, changing little during the coldest and warmest times when sunshine was nearly constant, but changing much during the times when the sunshine was changing. This leads to a second possible interpretation, that we should avoid rapid changes in greenhouse gases or other factors that could trigger even more rapid changes in climate. This is the "drunk" model of the climate system—when left alone, it sits; when forced to move, it staggers.

Certainly, the climate record does tell us that great warmth is not a guarantee of stability. As noted above, about 8,200 years ago a sharp cooling dropped central Greenland temperatures roughly 10°F for a century. Drying in Africa, increased windiness off Venezuela, cooling in Europe especially in the wintertime, and cooling of the surface of the north Atlantic indicate that this short-lived event was quite similar to the many older coolings of the Dansgaard-Oeschger oscillations. This event 8,200 years ago seems to have just followed the last major collapse of ice and ice-marginal lakes in Hudson Bay as the Canadian ice sheet died. The sudden flood of fresh water into the north Atlantic from this collapse probably shut down some of the conveyor belt circulation, triggering the shutdown pattern of climate changes. The significance of this most recent event is that, because of orbital changes, typical northern temperatures before and after the event were a bit higher than recent ones, especially during summers. So, a warm climate can jump if given enough reason.

Pushing the Drunk

Nature certainly can start the climate jumping again. But can humans? The answer is "maybe." At least some model results suggest that if humans warm the world too rapidly, increased rainfall and melting of Greenland ice and other glaciers in the far north will supply enough fresh water to the north Atlantic to stop the conveyor. The model results do not agree on how close we are to that threshold, however, or how rapidly the change will occur.

One particularly instructive model is that developed by Thomas Stocker of the University of Bern, Switzerland. Thomas has built a simple model, which includes the essential features of the oceans and atmosphere, and nothing more. Because his model is simple, he has the computer power to simulate changes over long periods of time and to simulate changes caused by many different things; really complex models tax the capabilities of even the world's fastest supercomputers, so they do not allow for a lot of experiments. Thomas has tested his model against the paleoclimatic record, and found that the model works well. He then has run the model into the future, and asked how different possible human behaviors would affect the conveyor circulation.

The model shows that if we raise greenhouse gases rapidly and greatly, we will kill the conveyor. But, if we raise greenhouse gases only a little, or if we raise greenhouse gases a lot but slowly, the conveyor will weaken but will then recover.

What might a conveyor shutdown mean to humanity? In part, the effect will depend on when this shutdown is assumed to happen. If greenhouse warming is large before the conveyor weakens much, then the carbon dioxide may provide some or all of the wintertime warmth that the north Atlantic region loses when the ocean circulation slows. If a shutdown were to happen soon, it could produce a large event, perhaps almost as large as the Younger Dryas, dropping northern temperatures and spreading droughts far larger than the changes that have affected humans through recorded history, and perhaps speeding warming in the far south. The end of humanity? No. An uncomfortable time for humanity? Very.

We must also remember the "ignorance clause" here. We don't know what is coming, we aren't advanced enough to predict the changes yet, there is no need to panic or run for the southern hemisphere—but we've learned a lot, we know that major events are possible, and we suspect that there are other surprises waiting out there somewhere in the climate system.

18

By now, we have extracted a lot of information from the Greenland ice cores and many other sources. I hope you are convinced that the climate has changed in the past—greatly, rapidly, and across much of Earth. Such changes could happen again, and cause grave problems for humans. Humans ourselves might trigger such changes. What should we do about this?

The simple answer is that I don't know. As a scientist, I am one of those lucky people who are paid to go to fascinating places with wonderful people and learn new things. We hope that the things we learn are of use to humanity. Certainly, as humans have accumulated and applied knowledge, we have gained the ability to do many things that we want. Balanced against the scares from toxic chemicals, weapons of mass destruction, global warming, and the other ills of technology, we have the wonders of medicine, construction, transportation, communication, and more. And on balance, the good is winning, no matter what your newspaper might have said this morning—we live longer, healthier, more comfortable lives than ever before.

But while knowledge is the source of much of the success of humans, and much of that knowledge has come from scientists, this does not mean that scientists have any special claims to wisdom in the application of knowledge. True, Benjamin Franklin, Thomas Jefferson, and other scientists gained knowledge and used that knowledge wisely, but there have been plenty of scientists who said or did boneheaded things in the real world. In all probability, a scientist is just about as good as any other informed person in deciding what we should do.

That having been said, I'm going to speculate about the future and what we might do to meet it. Just remember that some of what follows is not science—it is the opinion of one human being who convinced the editors and reviewers of an excellent university press to publish his manuscript.

Looking into my ice-crystal ball, I believe that:

1. The climate will change. All right, that's an easy one. We cannot find any long time in the past when Earth's climate was entirely stationary. The "boring" climate of our current warm period saw the droughts that hurt the Anasazi in the U.S. west, the cooling into the Little Ice Age that drove the Vikings from Greenland, the drying that turned "the land of milk and honey" into a forbiddingly arid place, the growth of the Sahara as great lakes were converted to sand dunes, and many other changes. Before our current "boring" period, the climate was even more changeable. Change is the only unchangeable reality, and change will continue.

2. Climate change will give winners and losers. Another easy one. If snowfall is reduced, the operators of ski areas and commercial snowplows will be unhappy, but people who hate driving on ice will be glad. Rain favors ducks, and drought favors cactus. There is nothing so extreme that someone won't like it, and nothing so desirable that someone won't be unhappy with it.

3. In the short term, losers from climate change will outnumber winners. This may not be quite so obvious, but it is a different way of saying that people are pretty smart. We build for the climate we have. We put in enough wells to supply water during dry times, enough dams to stop floods during wet times, enough

capacity to warm our buildings in the winter and cool them in the summer. A wetter climate may require more dams or moving houses off flood plains, a drier climate more wells or more water conservation, a hotter climate more air conditioners or more deodorant sticks, a colder climate more heaters or more sweaters. With infrastructure optimized for the current climate, any change will force us to stop doing some other things (stamping out poverty, curing disease, watching professional wrestling on TV, or whatever else we do with our time) so we can adapt to climate change instead.

4. In the long term, losers from climate change probably will outnumber winners. This is where my thoughts start to get contentious. I don't know this, and neither does anyone else. But history shows that the falls of many civilizations have occurred when climate was changing. A healthy civilization can probably handle a rather large climate change, but a civilization that is barely getting by has grave difficulties, and the extra stress of a changing climate can "push it over the edge." Despite the great advances made by humanity, billions of people are still struggling on the edge of poverty and just ahead of famine, or locked in wars, or otherwise balanced on the edge. The more developed countries probably can handle the coming changes, and perhaps some residents of the developed world will find that they like the new world better (although they are more likely to become nostalgic instead), but the weaker economies do not need any more stress.

5. Slowing down a little may help us a lot. An abrupt change is harder to deal with than a gradual one. Air conditioners and furnaces last only a few decades, so you can buy a bigger machine when replacement time comes if a climate change takes more than a few decades. For faster changes, you have to waste some money buying a new air conditioner even when the old one is not worn out. We have seen that the climate

may be a bit like a drunk—when left alone, it sits; when forced to move, it staggers. Thomas Stocker's models suggest that the rate of global warming is as important as the amount of global warming in determining whether the north Atlantic circulation will shut down or continue on. Notice the logical problem here, however. An economist usually doesn't worry about things very far in the future—the best way to deal with uncertainty in most economic models is to build the biggest possible economy, and let that economy deal with whatever happens. And in point 4, I noted that the countries that have built bigger economies in the past are the ones that are more likely to be able to handle the coming changes. Slowing down is an insurance policy; we just need to decide how much insurance we want. (Notice that only well-off people carry insurance policies. If you are desperately trying to find enough food for tomorrow, you probably aren't spending too much of your time worrying about food for next year. So insurance policies raise all sorts of first-world/third-world conflicts as to who should slow down, and by how much. Many wise heads will be required to sort this one out.)

6. Saving some excess capacity may make our lives much easier in the future. This is another insurance policy. Nature gives us land, water, air, plants, animals, nutrients, and more. If we use everything that nature gives us, and then nature takes some back, we will have nowhere else to go. Some of us may starve, others may just be rather hungry or thirsty. If we leave some excess capacity out there, then we will have additional resources on which to draw if we need them. And those resources can be used by the other species with which we share this globe and on which we often rely. Again, it is difficult to tell starving people not to eat what is in front of them, which brings us to the most contentious issue of all.

7. Too many people will use up our excess capacity. Certainly, clever people develop resources where, before, there was nothing useful. Sand is turned into computer chips, and computers figure out ways to cut heating bills, allow us to replace expensive and wasteful travel with cheap and low-impact communications, and otherwise help us conserve resources. Many people point to this trend and use catch phrases such as "Multiply resources, don't divide them" or "A rising tide lifts all ships." But more people still want more space, more water, and more air and food. Even as communications have improved, our urge to travel has gone up.

We can decrease the impact of each person on Earth, and so save excess capacity. Indeed, we are doing this. The DDT of Rachel Carson's *Silent Spring*, those freons that were rapidly eroding the ozone layer, and the phosphates that were turning Lake Erie into a scummy green pond have largely been cleaned up, or will be soon. The history of lead pollution in Greenland snow includes a barely perceptible blip from Roman activities, followed by a rise beginning with the Industrial Revolution and taking off with the widespread use of leaded gasoline after World War II, that threatened to dull our world with an insidious poison. But the subsequent cleanup activities have been surprisingly effective in lowering lead concentrations in our environment; we can reduce our impact greatly, as shown in Figure 18.1. Efficient farming practices allow us to feed more people on less land, setting some places aside for parks and wilderness areas. Extinctions are stealing our biodiversity, but much is still being preserved. And the rate of extinction per person has almost certainly dropped drastically.

This last item deserves a bit of explanation. As humans introduced themselves and their associated rats and dogs to small islands, the species unique to those islands suffered mass extinctions. These island extinctions had much to do with habitat destruction and with the inability of the spe-

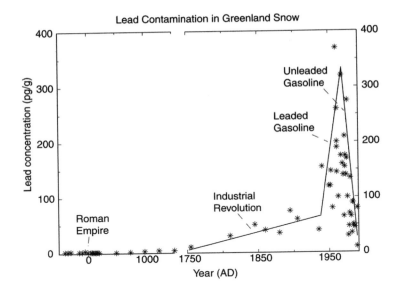

FIGURE 18.1

The history of lead concentration in Greenland snow with lines showing general trends, from the papers by Hong et al., and Boutron et al., cited in the Sources and Related Information. The tiny perturbation by the Romans is from smelting of "plumbum" for their plumbing. The huge rise in lead after World War II is from leaded automobile fuels; the precipitous fall then shows the effects of unleaded fuels and other changes made to clean up the environment. The units are trillionths of a gram of lead per gram of snow, which is picograms per gram or pg/g.

cialized types that lived on the islands to compete with the new generalists; it was not an intentional assault on species, but it happened. Also, at the end of the last ice age, the mammoths, mastodons, and numerous other large animals became extinct. Rapid climate changes have been blamed, but we now know that these large mammals lived through tens or hundreds of similar rapid climate changes before their recent extinction. The unique aspect of the most recent rapid climate change was that, as the shifting weather patterns stressed the remaining animals, they were also being stabbed to death by people with fluted-point spears.

Yet despite the drop in impact per person, our impact as

a species is going up. Almost anyone watching our surroundings knows that we are cutting, paving, plowing, burning, netting, or otherwise gathering more of nature's resources as our own. Virtually all indices of human impact show the same effect—we are almost everywhere, and increasingly using what is available. By some estimates, and in very crude numbers, we are using almost half the resources that Earth provides for our use and to share with all the other living things on the planet. Our numbers have been rising faster than our impact per person is dropping, so the share of Earth's bounty that we use is likely to keep going up.

So perhaps we will become so efficient that we can handle our growing numbers while saving some of the planet in a natural state. (There are still plenty of ways for us to lower our impact per person, especially for those of us living in the United States.) Perhaps we will use all our resources, and then suffer a terrible crash when the inevitable climate change takes some of those resources away from us. I hope for a soft landing, with our numbers stabilizing while our impact per person drops. I don't know how one will achieve such a soft landing—there are strong biological, religious, political, and personal factors involved. Improved health and education for all—so that people don't need extra babies as insurance in case the first ones die young, and so people have many choices for worthwhile things to do with their lives—may allow a soft landing without insulting those important religious, political, and personal beliefs. History shows that as people learn and grow, they tend to choose smaller families, not because someone told them to, but because they want to.

This last idea is a bit scary, because with education, people often learn about things that they then want, including cars and televisions and travel and health care. In the short term, these things will increase our impact per person. Do we want to risk this?

Suppose you need to jump your bicycle across a trench in the ground. If you ride up slowly, you'll fall in. But if you speed forward, you may gain the momentum to leap safely across. Humanity faces jumping across the trench formed as changing climates and environmental degradation undermine our ability to meet our own rising expectations. We cannot yet see into the trench to learn whether it is a shallow one that will inconvenience us, or a deep one that could swallow much of what we hold dear. Until we learn enough to see into the trench, prudence suggests that we should play it safe and prepare to jump. Educating and empowering people will speed us toward the edge, but should allow us to leap across to a soft landing on the other side.

We have come to a rest stop on the long road from central Greenland to our future. We know that humans have built a civilization adapted to the climate we have. Increasingly, humanity is using everything that this climate provides. Changes in the "normal" climate of the last few thousand years have had grave consequences for humans. But the Greenland ice cores and many other records show that the climate of the last few thousand years is about as good as it gets—most of the last 110,000 years have involved larger, faster, more-widespread climate changes.

The highway toward understanding climate change stretches off into the distance, running through the ice of Antarctica, the mountain glaciers of the Himalaya, the lakes of the Andes, the tree rings of the Arctic north, satellites that are probing the Earth system today, computer models, and the brains of people trying to grasp the whole picture. Our future will be much brighter if we can traverse this highway before we are overtaken by changing climate. Have a safe trip!

APPENDIX **I**

It is impossible to mention the hundreds of people who contributed to the success of GISP2, and the thousands who have worked on the ocean cores, tree rings, model building, and other aspects of this story. Yet it is equally impossible in good conscience to finish a book such as this without mentioning at least a few more of the central individuals, and apologizing to the many passed over.

The U.S. side of the push to collect the GISP2 ice core traces back to the 1985 report of the Committee for Science Planning in Greenland, chaired by Ellen Mosley-Thompson of Ohio State University and including Tony Gow. Wally Broecker and a distinguished consensus panel then did much to advance our cause.

The money side is easily overlooked but absolutely central, and the money for GISP2 came mostly from U.S. taxpayers through the National Science Foundation's Office of Polar Programs. The unsung heroes there provide diligent stewardship at minimal expense. Among many, Herman Zimmerman and Julie Palais deserve special mention for building the project and seeing it through to completion. A few of us received "extra" funding from other sources, including our universities and other government agencies, such as NASA. I also had a big boost from a fellowship from the David and Lucile Packard Foundation.

On the science side, GISP2 attempted to cover the major measurements, and to use two teams of investigators for each measurement whenever possible. Paul Mayewski of the Climate Change Research Center at the University of New Hampshire handled the major-ion chemistry, working with Sallie Whitlow, Dave Meeker, and others. Because Paul was also chief scientist and additionally responsible for the Science Management Office, he was busy indeed. As I noted in chapter 3, Paul was the "hero" who lived and breathed GISP2 through long years of trial and triumph, and the seemingly endless details that come with such a huge project. Success demands a hero, and we were fortunate to have one and to be able to follow his lead. We were also fortunate that Paul found top-notch people, including Mark Twickler, Michael Morrison, and Jen Putscher, to help manage the science.

Stable isotopes of ice were split among Pieter Grootes, Minze Stuiver, and Travis Saling of the University of Washington, and Jim White, Bruce Vaughn and Lisa Barlow of the University of Colorado. Investigations of windblown dust concentrations

193

and volcanic ash at the University of New Hampshire were be-gun by Julie Palais, who then switched to the National Science Foundation and left the work in the able hands of Greg Zielinski. Pierre Biscaye of Lamont-Doherty Earth Observatory added min-eralogic and isotopic characterization of the dust to learn where it came from, and Michael Ram of the University at Buffalo used laser-light scattering from the dust to count annual layers.

In support of the chemical and other studies, GISP2 with GRIP collaborators pioneered air-snow transfer studies coordi-nated by Jack Dibb of the University of New Hampshire with Cliff Davidson of Carnegie-Mellon. Byard Mosher of the Univer-sity of New Hampshire, Roger Bales of the University of Arizona, and Randy Borys of the Desert Research Institute worked on re-lated areas in air-snow transfer. You can't understand air-snow transfer if you don't know what the weather is like, so Chuck Stearns of the University of Wisconsin provided automatic weather stations.

One of the most interesting indicators of past changes is methanesulfonate. It can be traced back to some algae in the ocean and so records productivity there, and in the atmosphere it serves to nucleate cloud droplets and so affect precipitation and the reflectivity of Earth to sunlight. Eric Saltzman and Pai-Yei Whung of the University of Miami worked on the record and the air-snow transfer of this interesting chemical, and even worked with the Penn State crew tracking the formation of visible layers in the snow using the methanesulfonate as a tracer.

Interactions of the Earth system with space can be discerned from the cosmogenic isotopes. Kunihiko Nishiizumi, Robert Fin-kel, and James Arnold of the University of California-Berkeley and the Lawrence Livermore Laboratory used beryllium-10, alu-minum-26, and chlorine-36 to track changes in the sun's activity, Earth's magnetic field, and the history of the ice sheet. Ed Boyle of MIT with Rob Sherrell (now at Rutgers) tracked sources of lead and other trace metals in the core, looking at human-caused pollution, volcanoes, and meteorite impacts. The cosmogenic iso-tope carbon-14 can be used to learn past snow accumulation rates. Cosmic rays hitting the ice make much of its carbon-14. When snow accumulates rapidly, it is buried deeper than the penetration depth of cosmic rays before much carbon-14 is made; when accumulation is slow, much carbon-14 is made. De-vendra Lal of the Scripps Institution of Oceanography pioneered this technique and demonstrated excellent agreement with the traditional technique of flow-correcting annual layer thicknesses.

Characterization of the ice core involved the electrical properties, worked on by Kendrick Taylor of the Desert Research Institute. Other physical properties were handled by Tony Gow and Deb Meese with Bruce Elder of the U.S. Army Cold Regions Research and Engineering Laboratory, and with help from my group from Penn State and Joan Fitzpatrick from the United States Geological Survey. Deb Meese also joined the dating effort during the second year and then chaired the dating committee for the rest of the project, a huge job of counting layers, comparing counts among different workers, and smoothing ruffled feathers when some preliminary counts didn't agree quite as well as we had hoped.

The trapped gases in ice cores usually attract more attention than anything else, and GISP2 was well supplied with analysts. Michael Bender has pioneered one technique after another, first at Rhode Island and now at Princeton. The gravitational fractionation work that Todd Sowers (now at Penn State) did with Michael, followed by the work of Jeff Severinghaus (now at Scripps) with Michael, Todd, and Ed Brook (Washington State University), has provided some of the most startling results on the rapidity and size of climate changes. Todd Sowers, as a graduate student, was tasked with taking huge dewars of liquified gases to Greenland to use in sample extractions, while keeping the pilots happy about carrying things that, if not properly packed, could have been rather unsafe, and Todd did a brilliant job indeed. Martin Wahlen at Scripps and Wally Broecker at Lamont-Doherty Earth Observatory looked at carbon dioxide, and Alex Wilson and Douglas Donahue of the University of Arizona provided their unique perspective on carbon dioxide.

Another pioneering aspect of the central Greenland cores was full integration of ice-flow and ice-core record studies. Fortunately, we were guided by the early work of Steve Hodge from the U.S. Geological Survey and John Bolzan from Ohio State. Ed Waddington of the University of Washington conducted numerous model studies, and worked on the borehole logging and paleothermometry with Gary Clow of the U.S. Geological Survey.

To be truly useful, all of these data have to be made available to others. Numerous workers at the National Snow and Ice Data Center, led by Roger Barry and Richard Armstrong, were fundamental in collecting, archiving, and distributing data. Some of those data can be accessed at *http://www.ngdc.noaa.gov/paleo/ icecore/greenland/summit/index.html* for the National Geophysical Data Center, and *http://www.gisp2.sr.unh.edu/GISP2* for the GISP2 home page maintained by the Science Management Office, University of New Hampshire. The large amount of ice remaining for future studies is carefully archived at the National Ice Core Laboratory, under the watchful eyes of Technical Director Joan Fitzpatrick and Curator Geoff Hargreaves, *http://nicl.cr.usgs.gov.*

And on the other side of the Atlantic, a similarly long and distinguished list of investigators was working on similar topics on the GRIP core. Without providing a full summary of the European investigators, I will simply note here that I have been honored to publish papers with more than a dozen of our GRIP colleagues. I must especially mention the intercomparison of the physical properties of the cores, which involved Tony Gow, Deb Meese and me meeting with Sigfus Johnsen (University of Copenhagen and University of Iceland) and Sepp Kipfstuhl and Thorsteinn Thorsteinsson (Alfred-Wegener Institute, Germany) to examine both cores. Also notable was work with Dorthe Dahl-Jensen (University of Copenhagen) and Hitoshi Shoji (Kitami Institute of Technology, Japan) on deformation of ice, and with Jean Jouzel (Laboratoire des Sciences du Climat et de l'Environnement, France), David Peel (British Antarctic Survey), and others on the meaning of the stable isotopes of ice. The opportunity to work with these and other researchers from Europe and around the world was especially valuable and useful.

One of the great privileges of a professor is to help fuel students and younger colleagues so they can pull into the intellectual passing lane and head for the horizon. One of the great burdens of a professor is to see how easily these new scientists pass their professors on the way to the horizon. I certainly went through this privilege and burden with the many Penn Staters involved in GISP2.

The foundations of ice-core dating using visible stratigraphy rest on the unique relationship between summer sunshine and changes in snow structure that produce hoar frost. Christopher Shuman, a postdoctoral researcher with me, developed a remarkably clever way to track hoar formation using polarization ratios of passive microwaves sensed by satellites. He documented that hoar-forming events occur uniquely during summertime and encompass all of central Greenland, solidifying their utility in dating.

Christopher also did wonderful visible stratigraphy on the core. After I had spent six weeks counting layers in 1992, a health emergency made it impossible for the planned person to cover the next six weeks. Chris graciously agreed to attempt the job, and to try to learn it during the brief days when the heavy-lift aircraft were shuttling between Sondrestrom on the coast and GISP2 before they returned to New York. This would allow me to go home to see my family. (At the end of GISP2, I had spent more than 10 percent of my younger daughter's life on the ice sheet.) Flying to GISP2 involves going from sea level to two miles up, and almost everyone gets at least a little altitude sickness. The first day Chris tried to study the cores, he was really more interested in standing up without getting sick. Fortunately, he recovers from altitude sickness faster than I do, and he managed a crash course the second day and then did a brilliant job for the rest of the season. I went back and "checked" some of his work during a later year, and decided that I was better off believing him than me.

Sridhar Anandakrishnan is best known as an Antarctic seismologist, but he advanced ice-core analyses in several ways while conducting research with me at Penn State. The sonic-velocity shear-wave technique he developed in the field was quite useful and is one of the great improvisiations of all time. His insights to inversion and statistical studies proved central in turning borehole temperatures into surface temperature history, teasing out the effect of changing snow accumulation on impurity loadings in ice, and learning about summertime temperature changes from the rare, thin melt layers in the core.

Peter Fawcett never made it to Greenland, but as a post-doctoral researcher, he and Ph.D. student Anna Maria Ágústsdóttir explained the unexpected relationship between ice isotopes and temperature as resulting from changes in the season of the year when most of the snow fell. They also provided key insights to the spatial patterns and propagation mechanisms of the abrupt climate changes.

Kurt Cuffey started on GISP2 as a Penn State undergraduate. His early work included the temperature measurements that showed how the sun causes depth hoar, a heroic mission involving round-the-clock, every-hour-or-two sampling that, for technical reasons, could not be automated. He also worked on visible stratigraphy of the core for dating, and other physical studies. Next, Kurt demonstrated the techniques for combining borehole temperatures and ice isotopes to learn past surface temperatures, adapting Sridhar's computational methods for the unique circumstances of Greenland. After he left us to move to the University of Washington for graduate studies, Kurt worked with Ed Waddington and Gary Clow to show that the surface of Greenland during the ice age was at times as much as 40°F colder than today.

Mark Fischer, for his Ph.D., provided fundamental insights to the physical properties of ice. And Sridhar, Peter, Kurt, and Mark have gone on to teaching faculties at other universities, while Chris is on a research faculty and Anna Maria is in government service.

Other "winners" in the Penn State GISP2 effort include Wanda Kapsner, who, for her master's degree, showed that abrupt climate changes caused storm tracks to shift during major reorganizations of the atmospheric circulation. Bachelor's student Glenn Spinelli made important analyses of the statistics of visible strata. Greg Woods, while still an undergraduate, was routinely required to give practical advice to Ph.D. scientists (including me) who couldn't figure out how to accomplish some difficult task on the ice. "Woody" also reached the point of not wearing gloves in the trench during core processing, because it wasn't cold, it was "only" 20-something below zero. Jim Sloan and Greg Jablunovsky did first-rate field work.

And, this list is far from complete. For example, it doesn't include Deb Detwiler, who accomplished the impossible office task of keeping all of this in order. If you consider that other groups at GISP2 relied on similarly long lists of invaluable people, you should begin to appreciate that this was indeed "big science".

You also may appreciate the educational role of the ice cores. Greg Jablunovsky, for example, went straight from measuring the sound velocity in ice cores to using the speed of sound to look for cracks in the wings of Air Force planes before those cracks could cause crashes. Dozens of similar stories can be told for students at other schools. Many of our colleagues today—Eric Steig at the University of Pennsylvania, Karl Kreutz at Woods Hole Oceanographic Institution, Joe McConnell and Greg Lamorey at the Desert Research Institute, and many others—gained at least some of their education on GISP2. In myriad ways, projects such as GISP2 or GRIP or the ongoing ice-core studies provide training grounds for the leaders of tomorrow.

APPENDIX **2**

In 1999, NASA lost a spacecraft on Mars and a lot of taxpayer money because of a miscommunication over units. Scientists and other technical people in the United States live in a peculiar world in which miles and meters, pounds and pascals mix in a confusing way. Our lives would be easier if the United States joined the rest of the world in using the "international system" (S. I. in French), also known as the metric system, rather than the "customary system" we use based on the old English system.

When I started teaching large classes at Penn State, I was fired by the opportunity to reach hundreds of students at a time, and I taught only in metric. But in talking to students and grading papers, I came to realize that many of my students didn't know what I was talking about. Certainly, the students had been exposed to the metric system, but the metric units were not familiar enough for some of these students to make conversions and keep up with conversations at the same time. I also found that non-U.S. students often could make the conversions readily—these students had learned the units when they learned English.

So I was faced with choosing what was more important—information on groundwater pollution and global change and biodiversity and geologic hazards, or information on converting kilometers to miles. It took me a few years, but I decided that global change and biodiversity are more important, and I switched back to using customary units. I still try to sneak in some conversion factors, and I have a conversion exercise in the back of the text I wrote, but I teach using words that the students are likely to know.

I hope that many readers of this book are technically literate or are from beyond U.S. shores, and thus know metric units. But I also hope that some of the students in my classes, and in similar classes across the United States, will pick up a book such as this. So I have used customary units in the book. (Besides, *The Two-Mile Time Machine* does sound better than *The Three-Kilometer Time Machine*.)

For conversions, about 25 millimeters = 1 inch, 1.6 kilometers = 1 mile, 0.3 meters = 1 foot; or, 1 millimeter = 1/25 inch, 0.6 mile = 1 kilometer, and 1 m is just over 3 feet. On Earth's surface, 1 kilogram = 2.2 pounds or 1 pound is just under 1/2 kilogram, and a metric ton (1,000 kilograms) is not too different from a customary ton (2,000 pounds). For temperature, a Celsius degree is 1.8 times larger than a Fahrenheit degree, -40 is -40 in Fahrenheit or Celsius, and is darn cold in either if you forget your gloves.

The information in this book was collected from literally hundreds of sources, mostly in the refereed scientific literature. Leading journals that publish climate-change topics include *Nature, Science, Paleoceanography, Geophysical Research Letters, Journal of Geophysical Research, Geology, Quaternary Science Reviews, Tellus*, and many others.

In this section, I will list some useful sources for additional information, by chapter. I will lean heavily on articles that are likely to be widely available, including sources beyond the refereed scientific literature, such as books, articles in popular scientific publications, and some web sites. I hope to provide enough information that a reader can find out where things came from, but in no way am I trying to provide the full level of references expected in the scientific literature, a task that would double the length of the book.

I have tried to make these notes accurate and up-to-date, but I offer no guarantees, especially as web sites, addresses and other contact information can change over time.

∎ FAST FORWARD

The few-year jumping of climate that occurs near major climate transitions is described by K. C. Taylor, G. W. Lamorey, G. A. Doyle, R. B. Alley, P. M. Grootes, P. A. Mayewski, J. W. C. White, and L. K. Barlow, 1993, "The 'Flickering Switch' of Late Pleistocene Climate change," *Nature*, v. 361, pp. 432–436.

The story of the Norse settlement in Greenland is told in L. K. Barlow, J. P. Sadler, A. E. J. Ogilvie, P. C. Buckland, T. Amorosi, J. H. Ingimundarson, P. Skidmore, A. J. Dugmore, and T. H. McGovern, 1997, "Interdisciplinary Investigations of the End of the Norse Western Settlement in Greenland," *Holocene*, v. 7, pp. 489–499.

Little Ice Age changes are also detailed in J. M. Grove, 1988, *The Little Ice Age*, London, New York: Methuen, 498 pp.

An ice-core record of interactions between humans and climate is discussed in L. G. Thompson, M. E. Davis, E. Mosley-Thompson, and K-b. Liu, 1988, "Pre-Incan Agricultural Activity Recorded in Dust Layers in Two Tropical Ice Cores," *Nature*, v. 336, pp. 763–765.

For human-climate interactions, also see D. A. Hodell, J. H. Curtis, and M. Brenner, 1995, "Possible Role of Climate in the Collapse of Classic Maya Civilization," *Nature*, v. 375, pp. 391–394.

The data in Figures 1.1 and 1.2 are from K. M. Cuffey and G. D. Clow, 1997, "Temperature, Accumulation, and Ice Sheet Elevation in Central Greenland through the Last Deglacial Transition," *Journal of Geophysical Research*, v. 102(C12), pp. 26,383–26,396.

One of the best treatments of the "Operator's Manual" for planet Earth is the introductory material published for the U.S. NASA Earth Observing System (EOS) and their subsequent science plan, which is available (among other places), at the web site *http://eospso.gsfc.nasa.gov/sci—plan/chapters.html*. Other interesting materials from NASA are at *http://www.earth.nasa.gov/*.

A summary of much recent research on paleoclimatic records of climate change was assembled by Past Global Changes (PAGES), a core project of the International Geosphere-Biosphere Program (*http://www.pages.unibe.ch/*), and is included in a special issue, 2000, *Quaternary Science Reviews*, v. 19, pp. 1–479.

2 POINTERS TO THE PAST

A good textbook overview of how one can "read" sediments to learn about past climates is provided in R. S. Bradley, 1999, *Paleoclimatology: reconstructing climates of the Quaternary, 2nd ed.*; San Diego, CA, London: Academic Press, 613 pp.

Shorter overviews focused on the ice-core record include R. B. Alley and M. L. Bender, 1998, "Greenland Ice Cores: Frozen in Time," *Scientific American*, v. 278, pp. 80–85; and, K. Taylor, 1999, "Rapid Climate Change," *American Scientist*, v. 87, pp. 320–327.

An outstanding, accessible academic overview of much of the material covered in this book is provided by W. S. Broecker, 1995, *The Glacial World According to Wally, 2nd ed.*, Eldigio Press. This is basically a self-publication, but W. S. Broecker is so well known and respected that he can do this successfully. The latest contact information I have for obtaining a copy of this book is through Patty Catanzaro, Eldigio Press, Lamont-Doherty Earth Observatory of Columbia University, Palisades, NY 10964, USA; tel. 914-365-8515; fax 914-365-8155; e-mail *pcat@ldeo.columbia.edu*. Shorter treatments of much of the material from *The Glacial World According to Wally*, and of much of the material in *The Two-Mile Time Machine*, are given in W. S. Broecker, 1995, "Chaotic Climate," *Scientific American*, v. 273, pp. 44–50; and W. S. Broecker and G. H. Denton, 1990, "What Drives Glacial Cycles?", *Scientific American*, v. 262, pp. 49–56.

3 GOING TO GREENLAND

The U.S. Army Cold Regions Research and Engineering Lab was previously the Snow, Ice and Permafrost Research Establishment. The lab has, for some decades, been at 72 Lyme Road, Hanover, NH 03755, USA, telephone 603-646-4100, and the web site *http:// www.crrel.usace.army.mil.*

The career of Henri Bader and his pivotal role in ice coring is described very briefly in M. de Quervain and H. Rothlisberger, 1999, "Henri Bader (1907–1998)," *Ice (News Bulletin of the International Glaciological Society)*, No. 120, 2d issue, pp. 20–22.

Many of the scientific results from the Dye 3 ice core, southern Greenland, are collected in C. C. Langway Jr., H. Oeschger, and W. Dansgaard, eds., *Greenland Ice Core: Geophysics, Geochemistry, and the Environment*, Washington, DC: American Geophysical Union, 118 pp.

The GISP2 deep ice coring project was run by the Science Management Office (SMO), Climate Change Research Center, University of New Hampshire, under the direction of chief scientist Paul Mayewski. A history of GISP2, a list of the principal investigators, and other information are archived on the SMO web page at *http://www.gisp2.sr.unh.edu/GISP2/.*

The National Ice Core Laboratory (NICL) is a joint project of the National Science Foundation and the U.S. Geological Survey, with the University of New Hampshire Climate Change Research Center as an academic partner and running the NICL Science Management Office. The laboratory is located in the Denver Federal Center, and on the web at *http://nicl.usgs.gov* or *http:// www.nicl-smo.sr.unh.edu/NICL.*

The Polar Ice Coring Office (PICO) is funded by the National Science Foundation Office of Polar Programs; when I was writing this, PICO was at the University of Nebraska-Lincoln under the direction of Karl Kuivinen, but the contract was up for bid and it was unclear where the office would be for the next five years.

4 THE ICY ARCHIVES

Sea-level change is covered in chapter 7 of J. T. Houghton, L. G. Meira Filho, B. A. Callander, N. Harris, A. Kattenberg, and K. Maskell, eds., *Climate Change 1995: The Science of Climate Change*, Cambridge University Press, 572 pp. (which actually has a 1996 publication date despite the title). This is *the* summary on climate change, and is relevant to all the chapters in *The Two-Mile Time Machine*.

For a textbook overview of glaciers and ice sheets and how they flow, see W. S. B. Paterson, 1994, *The Physics of Glaciers, 3rd ed.*, Oxford, England and Tarrytown, NY: Pergamon, 480 pp.; or R. LeB. Hooke, 1998, *Principles of Glacier Mechanics*, Upper Saddle River, N.J.: Prentice-Hall, 248 pp.

Original scientific papers on how ice flow has affected layer thicknesses in Greenland include R. B. Alley, D. A. Meese, C. A. Shuman, A. J. Gow, K. C. Taylor, P. M. Grootes, J. W. C. White, M. Ram, E. D. Waddington, P. A. Mayewski, and G. A. Zielinski, 1993, "Abrupt Increase in Snow Accumulation at the End of the Younger Dryas Event," *Nature*, v. 362, pp. 527–529; and K. M. Cuffey and G. D. Clow, 1997, "Temperature, Accumulation, and Ice Sheet Elevation in Central Greenland through the Last Deglacial Transition," *Journal of Geophysical Research*, v. 102(C12), pp. 26,383–26,396.

5 ICE AGE THROUGH THE ICE AGE

The longest tree-ring chronology is described in B. Becker, B. Kromer, and P. Trimborn, 1991, "A Stable-Isotope Tree-Ring Timescale of the Late Glacial/Holocene Boundary," *Nature*, v. 353, pp. 647–649.

Some other interesting tree-ring work is described by G. C. Jacoby, R. D. D'Arrigo, and J. Glenn, 1999, "Tree-Ring Indicators of Climatic Change at Northern Latitudes," *World Resource Review*, v. 11, pp. 21–29; and G. C. Wiles, P. E. Calkin, and G. C. Jacoby, 1996, "Tree-Ring Analysis and Quaternary Geology: Principles and Recent Applications," *Geomorphology*, v. 16, pp. 259–272.

A long annually layered ocean sediment record is described in K. A. Hughen, J. T. Overpeck, S. J. Lehman, M. Kashgarian, J. Southon, L. C. Peterson, R. Alley, and D. M. Sigman, 1998, "Deglacial Changes in Ocean Circulation from an Extended Radiocarbon Calibration," *Nature*, v. 391, pp. 65–68.

Ice-core dating by layer counting is described by D. A. Meese, A. J. Gow, R. B. Alley, G. A. Zielinski, P. M. Grootes, M. Ram, K. C. Taylor, P. A. Mayewski, and J. F. Bolzan, 1997, "The Greenland Ice Sheet Project 2 Depth-Age Scale: Methods and Results," *Journal of Geophysical Research*, v. 102(C12), pp. 26,411–26,423; and R. B. Alley, C. A. Shuman, D. A. Meese, A. J. Gow, K. C. Taylor, K. M. Cuffey, J. J. Fitzpatrick, P. M. Grootes, G. A. Zielinski, M. Ram, G. Spinelli, and B. Elder, 1997, "Visual-Stratigraphic Dating of the GISP2 Ice Core: Basis, Reproducibility, and Application," *Journal of Geophysical Research*, v. 102(C12), pp. 26,367–26,381. These papers also evaluate the errors involved, which have been small in Greenland, but are not zero.

The 1999 paper in *American Scientist* by K. Taylor (chapter 2 Sources) includes a picture of a snow pit.

The use of electrical conductivity measurements of ice cores to study climate change is described by K. Taylor, R. Alley, J. Fiacco, P. Grootes, G. Lamorey, P. Mayewski, and M. J. Spencer, 1992, "Ice-Core Dating and Chemistry by Direct-Current Electrical Conductivity," *Journal of Glaciology*, v. 38, pp. 325–332.

Identification of volcanic ash to confirm ages of ice is described by G. A. Zielinski, P. A. Mayewski, L. D. Meeker, K. Gronvold, M. S. Germani, S. Whitlow, M. S. Twickler, and K. Taylor, 1997, "Volcanic Aerosol Records and Tephrochronology of the Summit, Greenland, Ice Cores," *Journal of Geophysical Research*, v. 102(C12), pp. 26,625–26,640.

The Laki eruption in Iceland is discussed by R. J. Fiacco Jr., Th. Thordarson, M. S. Germani, S. Self, J. M. Palais, S. Whitlow, and P. M. Grootes, 1994, "Atmospheric Aerosol Loading and Transport Due to the 1783–84 Laki Eruption in Iceland, Interpreted from Ash Particles and Acidity in the GISP2 Ice Core," *Quaternary Research*, v. 42, pp. 231–240.

Independent ages of the end of the Younger Dryas cold event have been generated by numerous research teams; many are summarized by R. B. Alley, C. A. Shuman, D. A. Meese, A. J. Gow, K. C. Taylor, K. M. Cuffey, J. J. Fitzpatrick, P. M. Grootes, G. A. Zielinski, M. Ram, G. Spinelli and B. Elder, 1997, "Visual-Stratigraphic Dating of the GISP2 Ice Core: Basis, Reproducibility, and Application," *Journal of Geophysical Research*, v. 102(C12), pp. 26,367–26,381.

For clathrates in the GRIP core, see F. Pauer, S. Kipfstuhl, W. F. Kuhs and H. Shoji, 1999, Air clathrate crystals from the GRIP deep ice core, Greenland; a number-, size- and shape-distribution study, *Journal of Glaciology*, v. 45, p. 22–30.

6 HOW COLD OF OLD?

The use of ice-isotopic ratios to learn past temperatures is reviewed in the books by W. S. B. Paterson (chapter 4 Sources) and R. S. Bradley (chapter 2 Sources).

The classic paper on ice-isotopic ratios for paleothermometry is W. Dansgaard, 1964, "Stable Isotopes in Precipitation," *Tellus*, v. 16, pp. 436–468.

A major recent review of the utility of ice isotopes in paleothermometry is J. Jouzel, R. B. Alley, K. M. Cuffey, W. Dansgaard, P. Grootes, G. Hoffmann, S. J. Johnsen, R. D. Koster, D. Peel, C. A. Shuman, M. Stievenard, M. Stuiver, and J. White, 1997, "Validity of the Temperature Reconstruction from Water Isotopes in Ice Cores," *Journal of Geophysical Research*, v. 102(C12), pp. 26,471–26,487.

The use of borehole paleothermometry to help understand ice isotopes is discussed by K. M. Cuffey, R. B. Alley, P. M. Grootes, J. F. Bolzan, and S. Anandakrishnan, 1994, "Calibration of the $\delta^{18}O$ Isotopic Paleothermometer for Central Greenland, Using Borehole Temperatures," *Journal of Glaciology*, v. 40, pp. 341–349; by K. M. Cuffey, G. D. Clow, R. B. Alley, M. Stuiver, E. D. Waddington, and R. W. Saltus, 1995, "Large Arctic Temperature Change at the Glacial-Holocene Transition," *Science*, v. 270, pp. 455–458; and by S. J. Johnsen, D. Dahl-Jensen, W. Dansgaard, and N. Gundestrup, 1995, "Greenland Paleotemperatures Derived from GRIP Bore Hole Temperature and Ice Core Isotope Profiles," *Tellus*, v. 47B, pp. 624–629. Ice-isotopic ratios record atmospheric conditions, primarily at the level where clouds form, and borehole temperatures record near-surface conditions. There is no physical law requiring temperatures at these different elevations to vary together, but success of this Cuffey et al. method shows that the surface and cloud-level temperatures have varied together; had they not done so, the ice-isotopic ratios would not have been able to predict the borehole temperatures. It is likely that the surface temperatures varied more than did the temperatures aloft, though in the same direction, as described by the Cuffey et al. paper in *Science* in 1995.

Also of interest is direct reconstruction of past surface temperatures from borehole temperatures without using ice-isotopic ratios, as described by R. B. Alley and B. R. Koci, 1990, "Recent Warming in Central Greenland?", *Annals of Glaciology*, v. 14, pp. 6–8; and by D. Dahl-Jensen, K. Mosegaard, N. Gundestrup, G. D. Clow, S. J. Johnsen, A. W. Hansen, and N. Balling, 1998, "Past Temperatures Directly from the Greenland Ice Sheet," *Science*, v. 282, pp. 268–271, among many sources.

The explanation that the unexpected relationship between ice isotopes and temperature results from changes in the season when most snow fell is presented by P. J. Fawcett, A. M. Ágústsdóttir, R. B. Alley, and C. A. Shuman, 1997, "The Younger Dryas Termination and North Atlantic Deepwater Formation: Insights from Climate Model Simulations and Greenland Ice Core Data," *Paleoceanography*, v. 12, pp. 23–38. An alternate interpretation is given by E. A. Boyle, 1997, "Cool Tropical Temperatures Shift the Global Delta δ^{18}O-T Relationship: An Explanation for the Ice Core δ^{18}O-Borehole Thermometry Conflict?," *Geophysical Research Letters*, v. 24, pp. 273–276. It is highly likely that both mechanisms have contributed to the observed calibration.

7 DUST IN THE WIND

Chemical records of climate change in Greenland ice cores include P. A. Mayewski, L. D. Meeker, S. Whitlow, M. S. Twickler, M. C. Morrison, P. Bloomfield, G. C. Bond, R. B. Alley, A. J. Gow, P. M. Grootes, D. A. Meese, M. Ram, K. C. Taylor, and M. Wumkes, 1994, "Changes in Atmospheric Circulation and Ocean Ice Cover over the North Atlantic during the Last 41,000 Years," *Science*, v. 263, pp. 1747–1751; P. A. Mayewski, L. D. Meeker, M. S. Twickler, S. Whitlow, Q. Yang, W. B. Lyons, and M. Prentice, 1997, "Major Features and Forcing of High-Latitude Northern Hemisphere Atmospheric Circulation Using a 110,000-Year-Long Glaciochemical Series," *Journal of Geophysical Research*, v. 102(C12), pp. 26,345–26,366; and M. DeAngelis, J. P. Steffensen, M. Legrand, H. Clausen, and C. Hammer, 1997, "Primary Aerosol (Sea Salt and Soil Dust) Deposited in Greenland during the Last Climatic Cycle: Comparison with East Antarctic Records," *Journal of Geophysical Research*, v. 102(C12), pp. 26,681–26,698.

Cosmogenic isotopes in ice cores are described by S. Baumgartner, J. Beer, M. Suter, B. Dittrich-Hannen, H.-A. Synal, P. W. Kubik, C. Hammer, and S. Johnsen, 1997, "Chlorine 36 Fallout in the Summit Greenland Ice Core Project Ice Core," *Journal of Geophysical Research*, v. 102(C12), pp. 26,659–26,662; R. C. Finkel, and K. Nishiizumi, 1997, "Beryllium-10 Concentrations in the Greenland Ice Sheet Project 2 Ice Core from 3–40 ka," *Journal of Geophysical Research*, v. 102(C12), pp. 26,699–26,706; D. Lal, A. J. T. Jull, G. S. Burr, and D. J. Donahue, "Measurements of In Situ ^{14}C Concentrations in Greenland Ice Sheet Project 2 Ice Covering a 17-kyr Time Span: Implications to Ice Flow Dynamics," *Journal of Geophysical Research*, v. 102(C12), pp. 26,505–26,510; and L. R. McHargue and P. E. Damon, 1991, "The Global Beryllium-10 Cycle," *Reviews of Geophysics*, v. 29, pp. 141–158.

The mechanism by which meteorites are concentrated in special regions of Antarctica was presented by I. M. Whillans and W. A. Cassidy, 1983, "Catch a Falling Star: Meteorites and Old Ice," *Science*, v. 222, pp. 55–57.

Collection of micrometeorites in the South Pole water well is reported by S. Taylor, J. H. Lever, and R. P. Harvey, 1998, "Accretion Rate of Cosmic Spherules Measured at the South Pole," *Nature*, v. 392, pp. 899–903.

The complexity of how chemicals and dust are transferred from air to snow is covered in many of the papers listed above, and in the research book edited by E. W. Wolff and R. C. Bales, 1996, *Chemical exchange between the atmosphere and polar snow,* NATO ASI Series, Series I, Global Environmental Change, v. 43, Berlin: Springer-Verlag, 675 pp. One attempt at simplifying this is R. B. Alley, R. C. Finkel, K. Nishiizumi, S. Anandakrishnan, C. A. Shuman, G. R. Mershon, G. A. Zielinski, and P. A. Mayewski, 1995, "Changes in Continental and Sea-Salt Atmospheric Loadings in Central Greenland during the Most Recent Deglaciation," *Journal of Glaciology*, v. 41, pp. 503–514.

Chemical fingerprinting of dust to learn the source of the dust is described by P. E. Biscaye, F. E. Grousset, M. Revel, S. Van der Gaast, G. A. Zielinski, A. Vaars and G. Kukla, 1997, "Asian Provenance of Glacial Dust (Stage 2) in the Greenland Ice Sheet Project 2 Ice Core, Summit, Greenland," *Journal of Geophysical Research*, v. 102(C12), pp. 26,765–26,781.

Useful reviews of many aspects of ice cores are given by R. J. Delmas, 1992, "Environmental Information from Ice Cores," *Reviews of Geophysics*, v. 30, pp. 1–22; and M. Legrand and P. Mayewski, 1997, "Glaciochemistry of Polar Ice Cores: A Review," *Reviews of Geophysics*, v. 35, pp. 219–244.

8 TINY BUBBLES IN THE ICE

A good summary of the ice-core record of atmospheric composition is given by D. Raynaud, J. Jouzel, J. M. Barnola, J. Chappellaz, R. J. Delmas, and C. Lorius, 1993, "The Ice Record of Greenhouse Gases," *Science*, v. 259, pp. 926–933.

The history of human perturbations to the atmosphere is told by M. Battle, M. Bender, T. Sowers, P. P. Tans, H. H. Butler, J. W. Elkins, J. T. Ellis, T. Conway, N. Zhang, P. Lang, and A. D. Clarke, 1996, "Atmospheric Gas Concentrations over the Past

Century Measured in Air from Firn at the South Pole," *Nature*, v. 383, pp. 231–235.

The dramatic story of how temperature in Antarctica and greenhouse gas concentrations have changed over almost half a million years is given in J. R. Petit, J. Jouzel, D. Raynaud, N. I. Barkov, J. M. Barnola, I. Basile, M. Bender, J. Chappellaz, M. Davis, G. Delaygue, M. Delmotte, V. M. Kotlyakov, M. Legrand, V. Y. Lipenkov, C. Lorius, L. Pepin, C. Ritz, E. Saltzman, and M. Stievenard, 1999, "Climate and Atmospheric History of the Past 420,000 Years from the Vostok Ice Core, Antarctica," *Nature*, v. 399, pp. 429–436.

Ice-core correlations and climate interpretations based on ice-core gases are included in T. Sowers and M. Bender, 1995, "Climate Records Covering the Last Deglaciation," *Science*, v. 269, pp. 210–214; and L. G. Thompson, M. E. Davis, E. Mosley-Thompson, T. A. Sowers, K. A. Henderson, V. S. Zagorodnov, P. N. Lin, V. N. Mikhalenko, R. K. Campen, J. F. Bolzan, J. Cole-Dai, and B. Francou, 1998, "A 25,000-Year Tropical Climate History from Bolivian Ice Cores," *Science*, v. 282, pp. 1858–1864.

The effects of freons on ozone, global warming, and the like are treated in D. J. Hofmann, S. J. Oltmans, J. M. Harris, S. Solomon, T. Deshler, and B. J. Johnson, 1992, "Observation and Possible Causes of New Ozone Depletion in Antarctica in 1991," *Nature*, v. 359, pp. 283–287; S. Solomon and J. S. Daniel, 1996, "Impact of the Montreal Protocol and Its Amendments on the Rate of Change of Global Radiative Forcing," *Climatic Change*, v. 32, pp. 7–17; and S. Solomon, 1999, "Stratospheric Ozone Depletion: A Review of Concepts and History," *Reviews of Geophysics*, v. 37, pp. 275–316, among many good sources.

9 THE SAURIAN SAUNA

The faint-young-sun paradox and the long-term stability of Earth's climate are treated by J. C. G. Walker, P. B. Hays, and J. F. Kasting, 1981, "A Negative Feedback Mechanism for the Long-Term Stabilization of Earth's Surface Temperature," *Journal of Geophysical Research*, v. 86(C10), pp. 9776–9782; J. F. Kasting, 1989, Long-Term Stability of the Earth's Climate," *Palaeogeography, Palaeoclimatology, Palaeoecology (Global & Planetary Change Section)*, v. 75, pp. 83–95; and J. F. Kasting and D. H. Grinspoon, 1991, "The Faint Young Sun Problem," in C. P Sonnett et al., eds., *The Sun in Time*, Tucson, AZ: University of Arizona Press, pp. 447–462.

An excellent discussion of feedbacks, and how interactions among various feedbacks can greatly amplify small changes, is given by J. Hansen, A. Lacis, D. Rind, G. Russell, P. Stone, I. Fung, R. Ruedy, and J. Lerner, 1984, "Climate Sensitivity: Analysis of Feedback Mechanisms," in J. Hansen and T. Takahashi, eds., *Climate Processes and Climate Sensitivity*, Washington, DC: American Geophysical Union, pp. 130–163.

Effects of continental drift on climate are included in R. A. Berner, A. C. Lasaga, and R. M. Garrels, 1983, "The Carbonate-Silicate Geochemical Cycle and Its Effect on Atmospheric Carbon Dioxide over the Past 100 Million Years," *American Journal of Science*, v. 283, pp. 641–683; and L. A. Frakes, J. E. Francis, and J. I. Syktus, 1992, *Climate Modes of the Phanerozoic: The History of the Earth's Climate over the Past 600 Million Years*, Cambridge University Press, 274 pp.

Effects of meteorite impacts are discussed in O. B. Toon, K. Zahnle, D. Morrison, R. P. Turco, and C. Covey, 1997, "Environmental Perturbations Caused by the Impacts of Asteroids and Comets, *Reviews of Geophysics*, v. 35, pp. 41–78; with a popular account in W. Alvarez, 1997, *T. Rex and the Crater of Doom*, Princeton, NJ: Princeton University Press, 185 pp.

10 THE SOLAR SYSTEM SWING

The story of Milankovitch and orbital calculations and ice ages has been told many times. A surprising number of professionals learned the story from a popular book by J. Imbrie and K. Palmer Imbrie, 1979, *Ice Ages: Solving the Mystery*, Hillside, NJ: Enslow Publishers, 224 pp. The story is also told in W. S. Broecker, *The Glacial World According to Wally*, described in these Sources for chapter 2.

The size of ice-age temperature changes and the role of carbon dioxide in causing cooling are summarized by D. Pollard and S. L. Thompson, 1997, "Climate and Ice-Sheet Mass Balance at the Last Glacial Maximum from the GENESIS Version 2 Global Climate Model," *Quaternary Science Reviews*, v. 16, pp. 841–864.

The 100,000-year cycle in ice ages became dominant over faster cycles about one million years ago, which is also when the peaks in ice volume became bigger. A possible explanation of this, linked to the interaction of the north American ice sheet with soft sediments on the continent, is given by P. U. Clark and D. Pollard, 1998, "Origin of the Middle Pleistocene Transition by Ice Sheet Erosion of Regolith," *Paleoceanography*, v. 13, pp. 1–9.

Changes in land-surface type and area are covered by L. R. Kump and R. B. Alley, 1994, "Global Chemical Weathering on Glacial Timescales," In W. W. Hay, ed., *Material Fluxes on the Surface of the Earth*, New York: National Academy of Sciences, pp. 46–60.

The history of ice volume on Earth shown in Figure 10.1 is from J. Imbrie et al., 1990, SPECMAP Archive #1, IGBP PAGES/World Data Center-A for Paleoclimatology Data Contribution Series #90-001, NOAA/NGDC Paleoclimatology Program, Boulder, CO, *http://www.ngdc.noaa.gov:80/mgg/geology/specmap.html.* Original papers include J. Imbrie, A. McIntyre and A. C. Mix, 1989, "Oceanic Response to Orbital Forcing in the Late Quaternary: Observational and Experimental Strategies," in A. Berger, S. H. Schneider, and J.-C. Duplessy, eds., *Climate and Geo-Sciences: A Challenge for Science and Society in the 21st Century*, Dordrecht, Boston, London: Kluwer Academic Publishers, pp. 121–164.

11 DANCING TO THE ORBITAL BAND

The Sources and Related Information for chapter 10 cover much of the material in chapter 11 as well, and only a few additions are listed here.

A thorough technical treatment of the 100,000-year cycle is given in J. Imbrie, A. Berger, E. A. Boyle, S. C. Clemens, A. Duffy, W. R. Howard, G. Kukla, J. Kutzbach, D. G. Martinson, A. McIntyre, A. C. Mix, B. Molfino, J. J. Morley, L. C. Peterson, N. G. Pisias, W. L. Prell, M. E. Raymo, N. J. Shackleton, and J. R. Toggweiler, 1993, "On the Structure and Origin of Major Glaciation Cycles: 2. The 100,000-Year Cycle," *Paleoceanography*, v. 8, pp. 699–735.

An interesting perspective on this cycle is in P. U. Clark, R. B. Alley, and D. Pollard, 1999, "Northern Hemisphere Ice-Sheet Influences on Global Climate Change," *Science*, v. 286, pp. 1104–1111.

The data in Figures 11.1 and 11.2 on carbon dioxide concentration and Antarctic temperature are from J. R. Petit, J. Jouzel, D. Raynaud, N. I. Barkov, J. M. Barnola, I. Basile, M. Bender, J. Chappellaz, M. Davis, G. Delaygue, M. Delmotte, V. M. Kotlyakov, M. Legrand, V. Y. Lipenkov, C. Lorius, L. Pepin, C. Ritz, E. Saltzman, and M. Stievenard, 1999, "Climate and Atmospheric

History of the Past 420,000 Years from the Vostok Ice Core, Antarctica," *Nature*, v. 399, pp. 429–436.

Earth's carbon cycle and interactions with humans are explained in W. S. Broecker and T.-H. Peng, 1998, *Greenhouse Puzzles, Second Edition*, Eldigio Press, as described in these Sources for chapter 2 for *The Glacial World According to Wally*.

Another excellent source is J. T. Houghton, L. G. Meira Filho, B. A. Callander, N. Harris, A. Kattenberg and K. Maskell, eds., *Climate Change 1995: The Science of Climate Change*, Cambridge University Press, 572 pp. (which actually has a 1996 publication date despite the title).

12 WHAT THE WORMS TURNED

Details of the end of the Younger Dryas are found in K. C. Taylor, P. A. Mayewski, R. B. Alley, E. J. Brook, A. J. Gow, P. M. Grootes, D. A. Meese, E. S. Saltzman, J. P. Severinghaus, M. S. Twickler, J. W. C. White, S. Whitlow, and G. A. Zielinski, 1997, "The Holocene/Younger Dryas Transition Recorded at Summit, Greenland," *Science*, v. 278, pp. 825–827.

The gas-isotopic results were reported by J. P. Severinghaus, T. Sowers, E. J. Brook, R. B. Alley, and M. L. Bender, "Timing of Abrupt Climate Change at the End of the Younger Dryas Interval from Thermally Fractionated Gases in Polar Ice," *Nature*, v. 391, pp. 141–146.

Evidence on the involvement of low-latitude sites based on interhemispheric gradients in methane concentration is given by E. J. Brook, J. Severinghaus, S. Harder, and M. Bender, 1999, "Atmospheric Methane and Millennial Scale Climate Change," in P. U. Clark, R. S. Webb, and L. D. Keigwin, eds., *Mechanisms of Global Climate Change at Millennial Time Scales*, Washington, DC: American Geophysical Union, pp. 165–175.

Evidence for storm-track shifts over Greenland between warm and cold times is available in W. R. Kapsner, R. B. Alley, C. A. Shuman, S. Anandakrishnan, and P. M. Grootes, 1995, "Dominant Control of Atmospheric Circulation on Snow Accumulation in Central Greenland," *Nature*, v. 373, pp. 52–54.

A compilation of Younger Dryas information is D. M. Peteet, ed., 1993, "Global Younger Dryas," *Quaternary Science Reviews*, v. 12; also see R. B. Alley and P. U. Clark, 1999, "The Deglaciation of the Northern Hemisphere: A Global Perspective," *Annual Reviews of Earth and Planetary Sciences*, v. 27, pp. 149–182.

For changes in the Cariaco Basin, see K. A. Hughen, J. T. Overpeck, L. C. Peterson, and S. Trumbore, 1996, "Rapid Climate Changes in the Tropical Atlantic Region during the Last Deglaciation," *Nature*, v. 380, pp. 51–54.

The Figure 12.2 temperature record for the last 100,000 years in Greenland is from K. M. Cuffey and G. D. Clow, 1997, "Temperature, Accumulation, and Ice Sheet Elevation in Central Greenland through the Last Deglacial Transition," *Journal of Geophysical Research*, v. 102(C12), pp. 26,383–26,396.

Original papers on Dansgaard-Oeschger events are W. Dansgaard, S. J. Johnsen, H. B. Clausen, D. Dahl-Jensen, N. Gundestrup, C. U. Hammer, and H. Oeschger, 1984, "North Atlantic Climatic Oscillations Revealed by Deep Greenland Ice Cores," in J. Hansen and T. Takahashi, eds., *Climate Processes and Climate Sensitivity*, Washington, DC: American Geophysical Union, pp. 288–298; and H. Oeschger, J. Beer, U. Siegenthaler, B. Stauffer, W. Dansgaard, and C. C. Langway, 1984, "Late Glacial Climate History from Ice Cores," pp. 299–306 in the same volume.

A little information on the folding of ice cores is contained in R. B. Alley, A. J. Gow, S. J. Johnsen, J. Kipfstuhl, D. A. Meese, and Th. Thorsteinsson, 1995, "Comparison of Deep Ice Cores," *Nature*, v. 373, pp. 393–394. Some pictures of folds are buried in the technical report of R. B. Alley, A. J. Gow, D. A. Meese, J. J. Fitzpatrick, E. D. Waddington, and J. F. Bolzan, 1997, "Grain-Scale Processes, Folding, and Stratigraphic Disturbance in the GISP2 Ice Core," *Journal of Geophysical Research*, v. 102(C12), pp. 26,819–26,830.

The Bond cycle is first described in G. Bond, W. Broecker, S. Johnsen, J. McManus, L. Labeyrie, J. Jouzel, and G. Bonani, 1993, "Correlations between Climate Records from North Atlantic Sediments and Greenland Ice," *Nature*, v. 365, pp. 143–147.

Heinrich events are named for H. Heinrich, 1988, "Origin and Consequences of Cyclic Ice Rafting in the Northeast Atlantic Ocean during the Past 130,000 Years," *Quaternary Research*, v. 29, pp. 143–152. Another good source on Heinrich events is G. C. Bond, H. Heinrich, W. S. Broecker, L. D. Labeyrie, J. McManus, J. Andrews, S. Huon, R. Jantschik, S. Clasen, C. Simet, K. Tedesco, M. Klas, G. Bonani, and S. Ivy, 1992, "Evidence for Massive Discharges of Icebergs into the North Atlantic Ocean during the Last Glacial Period," *Nature*, v. 360, pp. 245–249. The importance of Heinrich events is emphasized in W. S. Broecker,

1994, "Massive Iceberg Discharges as Triggers for Global Climate Change," *Nature*, v. 372, pp. 421–424.

A model for ice-sheet surges causing the Heinrich events is presented in D. R. MacAyeal, 1993, "A Low-Order Model of Growth/Purge Oscillations of the Laurentide Ice Sheet," *Paleoceanography*, v. 8, pp. 767–773; D. R. MacAyeal, 1993, "Binge/Purge Oscillations of the Laurentide Ice Sheet as a Cause of the North Atlantic's Heinrich Events," *Paleoceanography*, v. 8, pp. 775–784; and R. B. Alley and D. R. MacAyeal, 1994, "Ice-Rafted Debris Associated with Binge/Purge Oscillations of the Laurentide Ice Sheet," *Paleoceanography*, v. 9, pp. 503–511.

A review of the evidence for climate changes beyond but tied to the north Atlantic is given by R. B. Alley and P. U. Clark, 1999, "The Deglaciation of the Northern Hemisphere: A Global Perspective," *Annual Reviews of Earth and Planetary Sciences*, v. 27, pp. 149–182.

Evidence that the pattern of climate changes from the Greenland ice cores was going on for longer times is contained in J. F. McManus, D. W. Oppo, and J. L. Cullen, 1999, "A 0.5-Million-Year Record of Millennial-Scale Climate Variability in the North Atlantic," *Science*, v. 283, pp. 971–975.

Evidence that the pattern of climate changes from the ice age has continued into the modern warm time includes L. D. Keigwin and G. A. Jones, 1989, "Glacial-Holocene Stratigraphy, Chronology, and Paleoceanographic Observations on Some North Atlantic Sediment Drifts," *Deep-Sea Research*, v. 36, pp. 845–867; and G. Bond, W. Showers, M. Cheseby, R. Lotty, P. Almasi, P. deMenocal, P. Priore, H. Cullen, I. Hajdas, and G. Bonani, 1997, "A Pervasive Millennial-Scale Cycle in North Atlantic Holocene and Glacial Climates," *Science,* v. 278, pp. 1257–1266.

13   HOW CLIMATE WORKS

The first part of this chapter is textbook material. I certainly am not a world expert on it; I got it from reading texts. The best one I know, although a little technical, is J. P. Peixoto and A. H. Oort, 1992, *Physics of Climate*, New York: American Institute of Physics. 520 pp. Also of interest is W. J. Schmitz Jr. and M. S. McCartney, 1993, "On the North Atlantic Circulation," *Reviews of Geophysics*, v. 31, pp. 29–50.

Introductions to the global conveyor circulation are given by W. S. Broecker, 1995, "Chaotic Climate," *Scientific American*, v.

273, pp. 44–50; and W. S. Broecker, 1997, "Thermohaline Circulation, the Achilles Heel of our Climate System: Will Man-Made CO_2 Upset the Current Balance?," *Science*, v. 278, pp. 1582–1588.

14 A CHAOTIC CONVEYOR?

Models showing the instability of the north Atlantic to freshwater delivery include T. F. Stocker, D. G. Wright, and W. S. Broecker, 1992, "The Influence of High-Latitude Surface Forcing on the Global Thermohaline Circulation," *Paleoceanography*, v. 7, pp. 529–541; S. Manabe and R. J. Stouffer, 1997, "Coupled Ocean-Atmosphere Model Response to Freshwater Input: Comparison to Younger Dryas Event," *Paleoceanography*, v. 12, pp. 321–336; and S. Rahmstorf, 1995, "Bifurcations of the Atlantic Thermohaline Circulation in Response to Changes in the Hydrological Cycle," *Nature*, v. 378, pp. 145–149.

Using ocean sediments to learn how ocean circulation has changed is described by two of the pioneers of this field, in E. A. Boyle, 1990, "Quaternary Deepwater Paleoceanography," *Science*, v. 249, pp. 863–870; and in *The Glacial World According to Wally* (see Sources for chapter 2).

Changes in ocean circulation based on such tracers have been reported in numerous papers, including L. D. Keigwin and S. J. Lehman, 1994, "Deep Circulation Change Linked to Heinrich Event 1 and Younger Dryas in a Middepth North Atlantic Core," *Paleoceanography*, v. 9, pp. 185–194; and M. Sarnthein, K. Winn, S. J. A. Jung, J. C. Duplessy, L. Labeyrie, H. Erlenkeuser, and G. Ganssen, 1994, "Changes in East Atlantic Deepwater Circulation over the Last 30,000 Years: Eight Time Slice Reconstructions," *Paleoceanography*, v. 9, pp. 209–267. This paper, along with R. B. Alley and P. U. Clark, 1999, "The Deglaciation of the Northern Hemisphere: A Global Perspective," *Annual Reviews of Earth and Planetary Sciences*, v. 27, pp. 149–182, and Stocker, T. F., 2000, "Past and Future Reorganizations in the Climate System," *Quaternary Science Reviews*, v. 19, pp. 301–319, have promoted the model of three modes of oceanic circulation.

The southern effects of the shutdown of the north Atlantic circulation have been modeled by T. F. Stocker, D. G. Wright, and W. S. Broecker, 1992, "The Influence of High-Latitude Surface Forcing on the Global Thermohaline Circulation," *Paleoceanography*, v. 7, pp. 529–541; T. J. Crowley, 1992, "North Atlantic Deep Water Cools the Southern Hemisphere," *Paleoceanography*, v. 7, pp. 489–497; and W. S. Broecker, 1998, "Paleocean Circulation during the Last Deglaciation: A Bipolar Seesaw?", *Paleoceanography*, v. 13, pp. 119–121. Data demonstrating that this effect occurs include T. Blunier, J. Chappellaz, J. Schwander, A. Dallenbach, B. Stauffer, T. F. Stocker, D. Raynaud, J. Jouzel, H. B. Clausen, C. U. Hammer, and S. J. Johnsen, 1998, "Asynchrony of Antarctic and Greenland Climate Change during the Last Glacial Period," *Nature*, v. 394, pp. 739–743; and other papers reviewed by R. B. Alley and P. U. Clark, 1999, "The Deglaciation of the Northern Hemisphere: A Global Perspective," *Annual Reviews of Earth and Planetary Sciences*, v. 27, pp. 149–182. An additional summary is provided in R. B. Alley, P. U. Clark, R. S. Webb, and L. D. Keigwin, 1999, "Making Sense of Millennial-Scale Climate Change," in P. U. Clark, R. S. Webb, and L. D. Keigwin, eds., *Mechanisms of Global Climate Change at Millennial Time Scales*, Washington, DC: American Geophysical Union, pp. 385–394.

Atmospheric transmission of the signal of north Atlantic shutdown has been modeled by many workers, including A. M. Ágústsdóttir, R. B. Alley, D. Pollard, and W. Peterson, 1999, "Ekman Transport and Upwelling from Wind Stress from GENESIS Climate Model Experiments with Variable North Atlantic Heat Convergence," *Geophysical Research Letters*, v. 26, pp. 1333–1336; and S. Hostetler, P. U. Clark, P. J. Bartlein, A. C. Mix, and N. G. Pisias, 1999, "Mechanisms for the Global Transmission and Registration of North Atlantic Heinrich Events," *Journal of Geophysical Research*, v. 104, pp. 3947–3952.

15 SHOVING THE SYSTEM

Volcanic effects on climate are considered by K. R. Briffa, P. D. Jones, F. H. Schweingruber, and T. J. Osborn, 1998, "Influence of Volcanic Eruptions on Northern Hemisphere Summer Temperature over the Past 600 Years," *Nature*, v. 393, pp. 450–455. For longer records of volcanism and climate, see M. Stuiver, P. M. Grootes, and T. F. Braziunas, 1995, "The GISP2 δ^{18}O-Climate Record of the Past 16,500 Years and the Role of the Sun, Ocean, and Volcanoes," *Quaternary Research*, v. 44, pp. 341–354; and G. A. Zielinski, P. A. Mayewski, L. D. Meeker, K. Gronvold, M. S. Germani, S. Whitlow, M. S. Twickler, and K. Taylor, 1997, "Volcanic Aerosol Records and Tephrochronology of the Summit, Greenland, Ice Cores," *Journal of Geophysical Research*, v. 102(C12), pp. 26,625–26,640.

Solar influences are discussed in D. Rind, J. Lean, and R. Healy, 1999, "Simulated Time-Dependent Climate Response to Solar Radiative Forcing since 1600," *Journal of Geophysical Research*, v. 104(D2), pp. 1973–1990.

The beryllium-10 record from the GISP2 ice core was discussed by R. C. Finkel and K. Nishiizumi, 1997, "Beryllium-10 Concentrations in the Greenland Ice Sheet Project 2 Ice Core from 3–40 ka," *Journal of Geophysical Research*, v. 102(C12), pp. 26,699–26,706. For the peak in beryllium-10 about 40,000 years ago and its relation to the magnetic field strength, also see F. Yiou, G. M. Raisbeck, S. Baumgartner, J. Beer, C. Hammer, S. Johnsen, J. Jouzel, P. W. Kubik, J. Lestringuez, M. Stievenard, M. Suter, and P. Yiou, 1997, "Beryllium-10 in the Greenland Ice Core Project Ice Core at Summit, Greenland," *Journal of Geophysical Research*, v. 102(C12), pp. 26,783–26,794; and A. Mazaud, C. Laj, and M. Bender, 1994, "A Geomagnetic Chronology for Antarctic Ice Accumulation," *Geophysical Research Letters*, v. 21, pp. 337–340.

Outburst floods affecting the north Atlantic and causing the Younger Dryas event have been studied by W. S. Broecker, M. Andree, W. Wolfli, H. Oeschger, G. Bonani. G. J. Kennett, and D. Peteet, 1988, "The Chronology of the Last Deglaciation: Implications to the Cause of the Younger Dryas Event," *Paleoceanography*, v. 3, pp. 1–19; and W. S. Broecker, J. P. Kennett, B. P. Flower, J. T. Teller, S. Trumbore, G. Bonani, and W. Woelfli, 1989, "Routing of Meltwater from the Laurentide Ice Sheet during the Younger Dryas Cold Episode," *Nature*, v. 341, pp. 318–321. The outburst-flood cause of the cold event 8,200 years ago is presented by D. C. Barber, A. Dyke, C. Hillaire-Marcel, A. E. Jennings, J. T. Andrews, M. W. Kerwin, B. Bilodeau, R. McNeely, J. Southon, M. D. Morehead, and J. M. Gagnon, 1999, "Forcing of the Cold Event of 8,200 Years Ago by Catastrophic Drainage of Laurentide Lakes," *Nature*, v. 400, pp. 344–348. Additional insights to outburst-flood causes of other climate changes are presented by J. M. Licciardi, J. T. Teller, and P. U. Clark, 1999, "Freshwater Routing by the Laurentide Ice Sheet during the Last Deglaciation," in P. U. Clark, R. S. Webb, and L. D. Keigwin, eds., *Mechanisms of Global Climate Change at Millennial Time Scales*, Washington, DC: American Geophysical Union, pp. 177–201.

How rapidly outburst floods supplied water to the oceans is a little difficult to constrain, but some floods probably equaled the world's biggest river for a while, and some may have equaled the sum of all of the world's rivers. Total water released often was similar to a large lake on Earth today, as detailed in some of the papers cited just above.

The oscillating north Atlantic model was developed by W. S. Broecker, G. Bond, and M. Klas, 1990, "A Salt Oscillator in the Glacial Atlantic? 1. The Concept," *Paleoceanography*, v. 5, pp. 469–477.

For information on El Niño and related processes, see M. A. Cane and S. E. Zebiak, 1989, "A Theory for El Niño and the Southern Oscillation," *Science*, v. 228, pp. 1085–1087; D. B. Enfield, 1989, "El Niño, Past and Present," *Reviews of Geophysics*, v. 27, pp. 159–187; S. G. Philander, 1990, *El Niño, La Niña, and the Southern Oscillation*, San Diego, California: Academic Press, 293 pp.; and D. E. Harrison and N. K. Larkin, 1998, "El Niño-Southern Oscillation Sea Surface Temperature and Wind Anomalies, 1946–1993," *Reviews of Geophysics*, v. 36, pp. 353–399; among other excellent sources. Reconstruction of El Niño history from tropical ice cores is covered in L. G. Thompson, E. Mosley-Thompson, and P. A. Thompson, 1992, "Reconstructing Interannual Climate Variability from Tropical and Subtropical Ice-Core Records," in H. F. Diaz and V. Markgraf, eds., *El Niño: Historical and Paleoclimatic Aspects of the Southern Oscillation*, Cambridge University Press, pp. 295–322.

Some information on processes in the Southern Ocean can be found in A. L. Gordon, B. Barnier, K. Speer, and L. Stramma, 1999, "Introduction to Special Section: World Ocean Circulation Experiment: South Atlantic Results," *Journal of Geophysical Research*, v. 104(C9), pp. 20,859–20,861 and in the papers that follow.

16 FUELISH

My favorite overview of the carbon cycle, human effects, and related subjects is by J. F. Kasting, 1998, "The Carbon Cycle, Climate, and the Long-Term Effects of Fossil Fuel Burning," *Consequences: The Nature and Implications of Environmental Change*, v. 4, pp. 15–27. Much valuable information is contained in W. S. Broecker and T.-H. Peng, 1998, *Greenhouse Puzzles, Second Edition*, published by Eldigio Press, as described in these Sources for chapter 2 for *The Glacial World According to Wally*. This book includes much information on iron fertilization of the ocean and its effect on carbon dioxide drawdown; also see W. S. Broecker, 1990, "Iron Deficiency Limits Phytoplankton Growth in Antarctic Waters; discussion," *Global Biogeochemical Cycles*, v. 4, pp. 3–4.

The "standard" source on global change from the United Nations-sanctioned Intergovernmental Panel on Climate Change (IPCC) is J. T. Houghton, L. G. Meira Filho, B. A. Callander, N. Harris, A. Kattenberg, and K. Maskell, eds., *Climate Change 1995: The Science of Climate Change*, Contribution of Working Group I to the Second Assessment Report of the Intergovernmental Panel on Climate Change, Cambridge University Press, 572 pp. (which actually has a 1996 publication date despite the title). The science is followed by R. T. Watson, M. C. Zinyowera, and R. H. Moss, eds., 1996, *Impacts, Adaptations and Mitigation of Climate Change: Scientific-Technical Analyses*, Contribution of Working Group II to the Second Assessment of the Intergovernmental Panel on Climate Change, Cambridge University Press, 878 pp.; and J. P. Bruce, H. Lee, and E. F. Haites, eds., 1996, *Economic and Social Dimensions of Climate Change*, Contribution of Working Group III to the Second Assessment of the Intergovernmental Panel on Climate Change, Cambridge University Press, 448 pp.

If you hang around the climate-change game long enough, you will hear someone complain about the IPCC, with the implication that it is some world-government cabal with designs on subverting national sovereignty in the the name of climate change, that it runs roughshod over the evidence, and that dissenting voices are suppressed. Based on my personal experience, such claims are a bunch of blather. I was invited to an IPCC writing workshop for the 1995 assessment. As nearly as I can tell, the invitation came because some of the important people in the process decided that a viewpoint was missing from the preparations (in regard to sea-level change from ice sheets), and they started down the list of people who might know enough to be able to write on that; eventually, they got to me, and I accepted the invitation. Most of the week was devoted to going over piles of review comments that had been collected from individuals, industry, organizations, and governments. The writing teams were required to take every comment seriously, and the lead authors made sure that we did so. The exercise is designed to produce an accurate view of the state of knowledge, and it does in the end focus on those ideas that have gained wide currency in the scientific community; it is not a perfect process, but I was impressed by the effort and the openness involved. I am playing some minor roles in the next assessment (contributing and reviewing), but I am not a major player—I just believe that the people involved are making an honest effort to do the best that they can. I don't believe in everything in the IPCC reports—who is ever happy with absolutely everything in any major project?—but I am favorably impressed with the IPCC process and product.

For a discussion of discounting and climate change, see W. D. Nordhaus, 1994, *Managing the Global Commons: The Economics of Climate Change*, Cambridge, Massachusetts: MIT Press; and P. A. Schultz and J. F. Kasting, 1997, "Optimal reductions in CO_2 emissions," *Energy Policy*, v. 25, pp. 491–500.

Also of interest is J. C. G. Walker and J. F. Kasting, 1992, "Effects of Fuel and Forest Conservation on Future Levels of Atmospheric Carbon Dioxide," *Palaeogeography, Palaeoclimatology, Palaeoecology (Global & Planetary Change Section)*, v. 97, pp. 151–189; and *Policy Implications of Greenhouse Warming: Mitigation, Adaptation, and the Science Base*, 1992, Panel on Policy Implications of Greenhouse Warming, National Academy of Sciences, National Academy of Engineering, Institute of Medicine, 944 pp.

A great deal of useful information can be found on the web sites of the U.S. Global Change Research Program, *http://www.usgcrp.gov/*, and of the IPCC, part of the United Nations Environment Programme, *http://www.ipcc.ch/*.

17 DOWN THE ROAD

The cooling 8,200 years ago is discussed by R. B. Alley, P. A. Mayewski, T. Sowers, M. Stuiver, K. C. Taylor, and P. U. Clark, 1997, "Holocene Climatic Instability: A Prominent, Widespread Event 8200 years ago," *Geology*, v. 25, pp. 483–486.

Evidence showing that there has been anomalous warming this century, and that we probably are "pushing the drunk," includes M. E. Mann, R. S. Bradley, and M. K. Hughes, 1998, "Global-Scale Temperature Patterns and Climate Forcing over the Past Six Centuries," *Nature*, v. 392, pp. 779–787; P. D. Jones, K. R. Briffa, T. P. Barnett, and S. F. B. Tett, 1998, "High-Resolution Palaeoclimatic Records for the Last Millennium: Interpretation, Integration and Comparison with General Circulation Model Control-Run Temperatures," *Holocene*, v. 8, pp. 455–471; M. E. Mann, R. S. Bradley and M. K. Hughes. 1999, "Northern Hemisphere Temperatures during the Past Millennium: Inferences, Uncertainties and Limitations," *Geophysical Research Letters*, v. 26, pp. 759–762; J. T. Overpeck, K. Hughen, D. Hardy, R. Bradley, R. Case, M. Douglas, B. Finney, K. Gajewski, G. Jacoby, A. Jennings, S. Lamoureux, A. Lasca, G. MacDonald, J. Moore, M. Retelle, S. Smith, A. Wolfe, and G. Zielinski, 1997, "Arctic Environmental

Change of the Last Four Centuries," *Science*, v. 278, pp. 1251–1256; H. N. Pollack, Shaopeng Huang and Po-Yu Shen, 1998, "Climate Change Record in Subsurface Temperatures: A Global Perspective," *Science*, v. 282, pp. 279–281; and J. Oerlemans, 1994, "Quantifying Global Warming from the Retreat of Glaciers," *Science*, v. 264, pp. 243–245.

Several papers cited in the Sources for chapter 14 bear on conveyor shutdown. Especially T. F. Stocker and A. Schmittner, "Influence of CO_2 Emission Rates on the Stability of the Thermohaline Circulation," *Nature*, v. 388, pp. 862–865, which suggests that the rate at which we release carbon dioxide is as important as how much we eventually release in the stability of the modern pattern of ocean circulation.

18 AN ICE-CORE VIEW OF THE FUTURE

Some interesting web sites on climate change include: *http://www.ngdc.noaa.gov:80/paleo/drought/drght—history.html*; *http://www.ngdc.noaa.gov:80/paleo/globalwarming*; and *http://www.al.noaa.gov/WWWHD/pubdocs/Assessment98.html.*

For a view of interactions of earlier humans with biodiversity, see E. O. Wilson, 1992, *The Diversity of Life*, New York: W. W. Norton, 424 pp.; D. Quammen, 1996, *The Song of the Dodo: Island Biogeography in an Age of Extinctions*, New York: Scribner, 702 pp.; and P. D. Ward, *The Call of Distant Mammoths: Why the Ice Age Mammals Disappeared*, New York: Copernicus, 241 pp.

The history of lead in Greenland snow is given in C. C. Patterson, C. Boutron, and R. Flegal, 1985, "Present Status and Future of Lead Studies in Polar Snow," in C. C. Langway Jr., H. Oeschger, and W. Dansgaard, eds., *Greenland Ice Core: Geophysics, Geochemistry, and the Environment*, Washington, DC: American Geophysical Union, pp. 101–104; C. F. Boutron, U. Goerlach, J.-P. Candelone, M. A. Bolshov, and R. J. Delmas, 1991, "Decrease in Anthropogenic Lead, Cadmium and Zinc in Greenland Snows since the Late 1960s," *Nature*, v. 353, pp. 153–156; and S. Hong, J.-P. Candelone, C. C. Patterson, and C. F. Boutron, 1994, "Greenland Ice Evidence of Hemispheric Lead Pollution Two Millennia Ago by Greek and Roman Civilizations," *Science*, v. 265, pp. 1841–1843.

Appendix I A CAST OF CHARACTERS

Some of the Penn State papers are cited for chapters above, including those by Ágústsdóttir, Cuffey, Fawcett, and Kapsner. Selected others include: S. Anandakrishnan, R. B. Alley, and E. D. Waddington, 1993, "Sensitivity of Ice-Divide Position in Greenland to Climate Change," *Geophysical Research Letters*, v. 21, pp. 441–444; S. Anandakrishnan, J. J. Fitzpatrick, R. B. Alley, A. J. Gow, and D. A. Meese, 1994, "Shear-Wave Detection of Asymmetric c-Axis Fabrics in the GISP2 Ice Core," *Journal of Glaciology*, v. 40, pp. 491–496; M. P. Fischer, R. B. Alley, and T. Engelder, 1995, "Fracture Toughness of Ice and Firn Determined from the Modified Ring Test," *Journal of Glaciology*, v. 41, pp. 383–394; C. A. Shuman and R. B. Alley, 1993, "Spatial and Temporal Characterization of Hoar Formation in Central Greenland Using SSM/I Brightness Temperatures," *Geophysical Research Letters*, v. 20, pp. 2643–2646; C. A. Shuman, R. B. Alley, S. Anandakrishnan, J. W. C. White, P. M. Grootes, and C. R. Stearns, 1995, "Temperature and Accumulation at the Greenland Summit: Comparison of High-Resolution Isotope Profiles and Satellite Passive Microwave Brightness Temperature Trends," *Journal of Geophysical Research*, v. 100(D5), pp. 9165–9177; G. Spinelli, 1996, "A Statistical Analysis of Ice-Accumulation Level and Variability in the GISP2 Ice Core and a Reexamination of the Age of the Termination of the Younger Dryas Cooling Episode," *Earth System Science Center Technical Report No. 96-001*, The Pennsylvania State University, University Park, PA; G. A. Woods, 1994, "Grain Growth Behavior of the GISP2 Ice Core from Central Greenland," *Earth System Science Center Technical Report No. 94-002*, The Pennsylvania State University, University Park, PA; and R. B. Alley and G. W. Woods, 1996, "Impurity Influence on Normal Grain Growth in the GISP2 Ice Core," *Journal of Glaciology*, v. 42, pp. 255–260.

Some of the fruits of European/Japanese/U.S. collaboration include: R. B. Alley, A. J. Gow, S. J. Johnsen, J. Kipfstuhl, D. A. Meese, and Th. Thorsteinsson, 1995, "Comparison of Deep Ice Cores," *Nature*, v. 373, pp. 393–394; D. J. Dahl-Jensen, T. Thorsteinsson, R. Alley, and H. Shoji, 1997, "Flow Properties of the Ice from the Greenland Ice Core Project Ice Core: The Reason for Folds?", *Journal of Geophysical Research*, v. 102(C12), pp. 26,831–26,840; and J. Jouzel, R. B. Alley, K. M. Cuffey, W. Dansgaard, P. Grootes, G. Hoffmann, S. J. Johnsen, R. D. Koster, D. Peel, C. A. Shuman, M. Stievenard, M. Stuiver, and J. White, 1997, "Validity of the Temperature Reconstruction from Water Isotopes in Ice Cores," *Journal of Geophysical Research*, v. 102(C12), pp. 26,471–26,487.

A writer takes credit for the words of a thousand people, and can only hope to have done justice to those from whom the words were borrowed. I've been fortunate to borrow from some of the best, and I thank them.

My family has been a source of strength and inspiration throughout. Many teachers of the Worthington, Ohio, public schools kept me on track (Nick Hainen gave me a knowledge of chemistry and a view of life that are as fresh now as they were in 1975). The geologists, geophysicists, and polar researchers at Ohio State and Wisconsin opened the world for me (I'm still working on problems, using tools, from Ian Whillans, and I'm still riding on the intellectual excitement from Charlie Bentley's program). Penn State has given me a home, great colleagues, and great students, without whom this book could never have happened (I still marvel that Eric Barron dared to hire me and give me the opportunities he did, and that the Geoscientists then embraced me as a member of the program).

I've been fortunate to work with some of the best ice-core analysts in the world on GISP2 and WAISCORES and other projects, with ice dynamicists through the West Antarctic Ice Sheet (WAIS) initiative, and with glacial geologists on the Matanuska Glacier and elsewhere. These are three of the finest communities of scientists anywhere, and I thank them.

Cindy Alley drafted most of the figures, and provided excellent feedback at many stages. Editors Kristin Gager and Jack Repcheck, production editors Ellen Foos and Jennifer Slater, and copy editor Deborah Wenger helped in many ways. Valuable discussions and reviews from Lisa Barlow, Michael Bender, Gerard Bond, Wally Broecker, Peter Clark, Joan Fitzpatrick, Tony Gow, Jim Kasting, Paul Mayewski, Deb Meese, Peter Schultz, Todd Sowers, Kendrick Taylor, Lonnie Thompson, and others helped me greatly at many stages in the writing.

I dedicate this book to my parents John and Ruth, my wife Cindy, and our daughters Janet and Karen, for giving me my past, my present and my future.

223